HISTORY OF THE FARMINGTON FIRE DEPARTMENT

1850 - 2000

A VOLUNTEER FIRE DEPARTMENT

BY ROBERT L. McCLEERY

FARMINGTON FIRE CHIEF 1977 - 2000

COMPILED, EDITED & UPDATED BY
RUTH McCLEERY WATSON

FOREWORD BY
DEPUTY CHIEF S. CLYDE ROSS

For permission contact:
Ruth McCleery Watson
PO Box 180
Kingfield, ME 04947

chiefdaughter477@gmail.com

Printed in the United States of America
First Printing, 2018

Ordering Information:
Contact the publisher at the address above or email chiefdaughter477@gmail.com

Publisher's Cataloging-in-Publication data
Author: McCleery, Robert L.
Title: History of the Farmington Fire Department 1850 - 2000/ A Volunteer Fire Department/
Robert L. McCleery; with Ruth McCleery Watson.

ISBN 978-0-692-11652-4

1. The main category of the book —History —Non-fiction. 2. Memoir - Non-fiction.
3. Biography - Non-Fiction.

Library of Congress Control Number: 2018904703

Disclaimer:
Although the author and publisher made every effort to ensure that the information
in this book was correct at press time, this work is the result of the author and publisher's compilation
of information from many and varied sources as well as the author's recollections of events, locales,
and conversations, as such the author and publisher do not assume,
and hereby disclaim, any liability to any party for any loss, damage, or disruption
caused by errors or omissions, whether such errors or omissions result from
negligence, accident, or any other cause.

The publisher would like to thank Cindy Butler for assistance with the graphic design
and printing of the final version of this book. She may be contacted at:
cindybutlerdesign@gmail.com

Typography: Mrs Eaves

Cover Photographs: 1992 Class "A" Burn Live Fire Training Session on Lincoln Street,
Chief Robert McCleery Pictured. Barbara Yeaton Photo.

Back Cover Photograph: Top: 1950 Farmington Fire Department Members at High Street Station.
Firemen Identification on PG. 78. Pictured: Chief Victor C. Huart, Chief Huart's successor,
J. Bauer Small, and Chief Small's successor, Forest L. Allen.
Below Left: Barrows Block Main Street Fire 12/15/1951. Below Right: Newman Motors Broadway Fire
4/03/1965. Photos Courtesy of Luce's Studio & Town of Farmington.

DEDICATION

This book is dedicated to the "boys" of the Farmington Fire Department past and present who have made the department what it is today. It has been my honor to work for and with the outstanding individuals of this department and my peers throughout the state. I would also like to dedicate this book to my wife, Edith, where it all starts, without her support and encouragement I would not have gotten very far in anything, and for her commitment of time and energy in support of this book and the Farmington Fire Department.

ACKNOWLEDGMENT

My sincere thanks go to the Town Managers, Board of Selectmen, and ladies in the office for their support over the years. I wish to thank the neighboring Fire Departments that so generously responded to our calls for mutual aid, the Farmington Police Department, Highway Department, dispatchers at the Dispatch Center, the Ladies Auxiliary for providing food and beverages when needed, the personnel of the former Hawthorne, Keegan, Delta and Life Star Ambulance crews for their assistance, and the citizens of Farmington who dedicate themselves in so many ways to this great town. My eternal thanks go out to all our fire fighters who worked with me every day to make the Farmington Fire Department one of the best in the state.

Robert L. McCleery

~ Robert L. McCleery

ABOUT THE BOOK

Our father passed away before he could finish this book. Dad spent many years researching the information contained herein recording it on stacks of yellow legal pads, and our mother spent untold hours typing his handwritten notes. After Dad's passing it was too much for our mother to complete without him so I have finally finished what Dad and Mom started. I have compiled as much information as possible that Dad was not able to source and have updated information to 2000 - 2017. It is with both joy and sadness that this project has reached completion; joy that Dad's work gets into the hands of you, the reader, and sadness that I may not feel as close to Dad as I have these past months reading his beautiful script, feeling his presence, and hearing his voice as I have worked on this project. Special thanks to Chief Terry S. Bell, Sr. and Deputy Chief S. Clyde Ross for their support and making themselves available to answer questions and gathering town annual reports and pictures; Thank you to Deputy Chief Tim Hardy for identifying firemen in pictures. I would also like to thank Farmington's picture archivist Don DeRoche, Town Clerk Leanne Dickey, and Farmington Historical Society Director Nancy Porter. Thank you to Town Manager Richard Davis for use of the conference room at the Municipal Building. Thank you to Curator Jim Orr at the Henry Ford Museum, Dearborn, MI for allowing the publication of the picture of the 1860 Hunneman Eagle fire engine; Archivist & Librarian Ethan Yankura of the Owls Head Transportation Museum, Owls Head, ME for permission to publish the 1916 Model T Chemical Fire Truck; the Town of Farmington for permission to use the annual reports; the newspaper publishers for allowing the use of their articles and photos. Thank you to Carrabassett Valley Library Director, Andrea DeBaise, and Prince Memorial Library (Cumberland, ME) Director, Thomas Bennett, for their suggestions. I would also like to thank my husband, Bill Watson, and our children, Chris, Will, Lane and Heath, for their love, support and encouragement; to friends and family members who have been supportive, I thank you. Please accept my apologies if I have missed anyone. This book contains Dad's personal recollections, observations and experiences. At times, there has been conflicting information, and at others, records were incomplete or lost as was the case of the early records of the Farmington Village Corporation. Information has been presented as accurately as possible and credit has been given where known. It is with heartfelt thanks that I thank all who have helped bring this project to fruition and you, the reader, for your interest in this book on the History of the Farmington Fire Department.

~ Ruth E. McCleery Watson

This plaque hung in our home on the wall behind Dad's favorite chair.

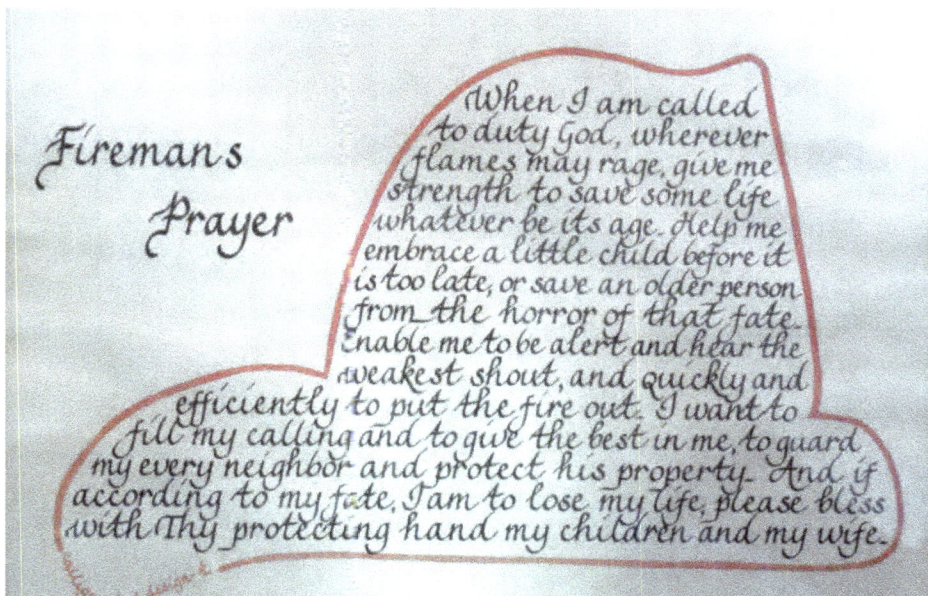

Fireman's Prayer

When I am called to duty God, wherever flames may rage, give me strength to save some life whatever be its age. Help me embrace a little child before it is too late, or save an older person from the horror of that fate. Enable me to be alert and hear the weakest shout, and quickly and efficiently to put the fire out. I want to fill my calling and to give the best in me, to guard my every neighbor and protect his property. And if according to my fate, I am to lose my life, please bless with Thy protecting hand my children and my wife.

By A.W. "Smokey" Linn, Wichita, KS Fireman 1947-1975.

Robert L. McCleery V

ABOUT THE AUTHOR

Upon Nomination for Fire Chief of the Year 1992 & 1998

"In my twenty-six plus years of local government experience, I have observed many local Fire Chiefs on a daily basis and none better than Chief McCleery."

—Town Manager John Edgerly, September 9, 1992

"Chief McCleery's extraordinary leadership has brought credit to his department, region and his profession. His willingness to lead by example and to dedicate endless hours to the enhancement and enrichment of the careers of firefighters throughout the state has been remarkable. Chief McCleery is revered by his department and recognized as a leader by his peers.

—Town Manager Pam Corrigan, September 14, 1998

Upon Announcement of his Retirement *Morning Sentinel Article*, Jan. 14, 2000

"People of his caliber are a rare breed to give that level of dedication - not only his professional life, but his personal life - in order to nurture the department."

—Selectman (and Firefighter) Stephen Bunker

"The citizens of Farmington are deeply indebted to Bob, and I hope they will take the opportunity to personally thank him."

—Selectman Charles Murray

"This is a sad milestone. The Chief has not only created a professional department here in Farmington, but has also bettered the professionalism of departments throughout the state by being an active participant in agencies involved with fire fighting."

—Town Manager Pam Corrigan

Upon Retirement July 2002

"Bob, as one who has been there, I understand the commitment that you have made to the department and your community. No one except another firefighter and the families of firefighters can appreciate what it means to be a firefighter and chief. And your 40 years of service is extraordinary and deserving of recognition. May you be blessed with a long and healthy life."

—State Fire Marshal John C. Dean, June 29, 2000

"Chief McCleery, your career spanning 40 years of service to the Farmington Fire Department and 23 years of chief is testimony to your lifelong dedication to your community. As you pass the role of leadership to your successor, you should feel a great sense of satisfaction, pride and confidence that you have provided outstanding guidance to those who will lead the department into the future. On behalf of the people of Maine, I wish you all the best in your retirement.

—*Governor Angus S. King, Jr., July 29, 2000*

Retired Farmington Fire Chief dies, *Sun Journal,* July 23, 2002

"Bob definitely made our department the way it is today…He was an easy person to work with and didn't mind sharing what he knew…Training was one of his big things." "He gave us the training we wanted and he was always open to new ideas and letting people learn."

—*Fire Chief Terry S. Bell, Sr.*

"It was McCleery's fairness and his encouragement to learn new skills that stands out in my mind…He always wanted us to be the best we could be. He was very, very proud of all of us…If it wasn't for Bob, I wouldn't be where I am today in the fire service. He was a great inspiration…He was an outstanding citizen who gave a lot to Farmington and the county."

—*Deputy Fire Chief Tim Hardy*

"I remember Bob for his honesty, his great sense of humor, and the fact that he was never afraid to tackle tough issues on the board."

—*Freeport Fire Chief Darryl Fournier, a friend and colleague for 20 years.*

(Both past presidents of the Maine Fire Chief's Association & board of directors members.)

"Bob was always on top of things, and he did it with a quiet, orderly manner, but when he was in a group of people he was often the ringleader in the conversation."

—*Former East Dixfield Fire Chief Richard "Dick" Hall, fellow farmer and friend.*

(Both served on the board of the Franklin County Agricultural Society for many years.)

"It was McCleery who helped me learn about the town when I became town manager in 1990. He was kind of a mentor. I'd go into talk with him and always got a good straight answer."

Former Town Manager John Edgerly

"Bob McCleery was a fire chief known to have a vision beyond his local area. He encouraged firefighters to seek knowledge outside the area, to bring back what they had learned and to adapt it to the Farmington Fire Department. 'Go ahead and try it boys, and we'll make it work' was one of his sayings...He didn't try to do it all...He was an on-call fire chief. He supported fire fighting training and increased training sessions locally, not only for his own fire fighters, but for those in other towns in Franklin County...He was a great advocate of self-contained breathing apparatus...He didn't just go out and buy a fire truck for today or tomorrow, he bought a fire truck for five, ten years down the road...Bob was a go-getter, with high expectations for his people and accountability of his people in the department."

—Deputy Fire Chief S. Clyde Ross

"McCleery was a leader among men...Highly regarded by those he led and peers."

—Chaplain Stanley Wheeler

In a Letter to the Editor of the Franklin Journal, August 16, 2002

"Every day his life had meaning, because every day he did things to benefit others, inspiring loyalty and love from all who knew him. Whether he was organizing events at the Franklin County Fair, raising livestock, hay or pumpkins, tending to children and grandchildren, or responding to late night fire and rescue emergencies, Chief McCleery served his fellow citizens of Farmington with tremendous commitment, endless energy and a great heart. Chief McCleery lived a fruitful purposeful life."

Maine State Attorney General Janet T. Mills

Other Comments

"I was so lucky to know your Dad. Your Dad was one of my favorites. He could fit in with anyone and was respected by all of us. He was certainly deserving of the Maine Fire Chief of the Year. He was a quiet leader and loved by all of us."

—Former Cape Elizabeth Fire Chief Philip McGouldrick, September 2, 2017

TABLE OF CONTENTS
NARRATIVE & CHRONOLOGY OF THE FARMINGTON FIRE DEPARTMENT

FOREWORD

The following accounts and happenings of the Farmington Fire Rescue Department are the personal observations of the former Chief Robert L. McCleery as taken from records accumulated over the years included in this manuscript. The presentation of important incidents and fires will offer one an opportunity to observe a number of significant changes in types, costs, equipment, and resulting effects of fires and accidents in our community and surrounding areas.

One should pay particular attention to the types of equipment purchased and the changes in equipment over the years. Costs, types, and numbers of apparatus are also mentioned in some detail. It is important to note that as our social environment advanced technically so did the demand for better fire fighting equipment, more responsible personnel, better training, and lastly adequate records. This paper reflects the evolution as seen through the eyes of one who pursued fire fighting for a number of years.

It is important to note the names of several dedicated leaders and personnel who have given of their time from work, family occasions, and other situations known only to them. The dedication cannot be measured in dollars, but in knowing that one has given of himself/herself to assist their community and fellow citizens or neighbors. Becoming a member of the fire department meant that one's life would never be the same after making this commitment. How this plays out is very basic, we live in a small town, know people personally, and have knowledge of the locations where emergency incidences have occurred.

In closing know that the reflections indicated here are those of one who has "walked the walk and talked the talk". One cannot change what has happened over the years, but we can note how our community has changed and fire protection has improved. If there is a misrepresentation of information contained herein, please note that documents often have conflicting accounts of how incidents have occurred. As you read try to imagine what conditions existed during the occasions presented.

~ S. Clyde Ross Deputy Chief Farmington Fire Rescue
Written September 1, 2017

S. Clyde Ross has been a member of the Farmington Fire Department for 46 years (5/4/71), promoted to Assistant Chief on February 6, 1978, ar d has held the position of Deputy Chief since August 6, 1991.

Chapter 1
OVERVIEW OF THE ORGANIZATION

In August of 1850, the Town of Farmington established the Farmington Village Corporation. It took nine years for the Village Corporation to organize the town's first fire department, and it wasn't until 1860 that the fire department was equipped and manned. Several major fires with significant property loss occurred during this period which roused some concerned individuals to demand that the Farmington Village Corporation appoint a committee to explore the cost of fire apparatus. A committee was formed, and it was voted to purchase an engine and other equipment, but nothing came of it because the community was unwilling to tax itself for fire protection. It took several more fires, with heavy property loss, to motivate the citizens to action. In June 1859, an act to incorporate the Farmington Village Corporation was accepted, and regulations were established for a fire department. A volunteer fire company was officially organized in 1860, and the town's first fire engine, "Eagle No. 1", a hand pumped fire engine, arrived on the 29th of November 1860.

The first fire house was located on Academy Street at the rear of what is now the administration building of the University of Maine in Farmington (formerly the Normal School and Farmington Academy). In 1864 when the Normal School took over the Academy property, the Fire Engine Company was moved to Pleasant Street, the former site of Stoyell's Shop, and the paint shop for the Newman Motor Company near the intersection of Front Street, Pleasant Street, and Broadway.

After the Great Fire of October 1886 there was a special town meeting on November 13, 1886 in which the town voted to purchase a steam engine and appliances to equip it at a cost of $5,000.00. In 1887, Farmington's steam fire engine arrived, and the fire department became the known as Farmington Steam Fire Engine Company No. 1 which had a steam fire engine, ladder wagon and hose carts. The fire house was moved to a building adjacent to the old Steam Laundry and the former showroom for J.W. & W.D. Barker, Inc. along High Street between Broadway and Church Street. At the annual town meeting held January 30, 1912, the Farmington Village Corporation adopted a charter and by-laws with rules and regulations for several departments. Farmington Fire Department No. 1 Rules and Regulations were established along with Constitution and By-Laws for the fire department, Special Duties of Officers and Others, and the Fire Alarm System. *See Chapters 15 & 16.*

Quote from the Farmington Village Corporation 1912 Charter and Bylaws: "Under the Charter of the Farmington Village Corporation approved by the Governor, March 20, 1911 and adopted by the Farmington Village Corporation, January 18, 1912, of an act to organize and maintain an efficient Fire Department, and to adopt all rules and regulations for the governing same, (ref. Chapter 142, Private and Special Laws of 1911) ...The name and title of this association shall be the

Farmington Fire Department Company No. 1...The purpose of the Department is to extinguish fires and to protect property in the Town of Farmington...The members shall be residents of the Farmington Village Corporation and of good moral character and temperate habits."

Article 2 - Duties of the Officers:

"The officers of the fire company shall consist of a Foreman who shall be the Chief Engineer of this department and appointed by the assessors. Other officers are First and Second Assistant Foreman, Clerk, Treasurer, and a Finance Committee of three, one of whom shall be the First Assistant Foreman. These officers shall be elected by ballot at the Annual Meeting held on the first Tuesday of December. The company shall hold regular monthly meetings on the first Tuesday of each month at 7:30 PM. Other officers appointed by the Foreman: Engineer and Assistants, Fireman, Foreman of Hooks and Ladders, First and Second Assistants, two Pipe men (nozzle) for each line, two Assistant Pipe men, two Hydrant men, two or more men to adjust suction hose, one lineman who shall be appointed a Special Police by the Assessors."

In the late 1800's and early 1900's, fire bells were used to alert the fire company to report for a fire; prior to that church bells most likely sounded the call. On July 26, 1900, a fire started around 5:00 PM at the A. H. Abbott property consuming the two main buildings on the property: A. H. Abbott family's valuable library and the Little Blue School for Boys. The fire alarm bells used to alert the firemen did not work, or malfunctioned, causing a delay in calling out the fire department. This disastrous fire, and alarm bell malfunction, led to a change in the town's fire alarm system resulting in the installation of an audible air horn in 1912 on the new fire station that had been built on the corner of High and Perham Streets, across from the Octagon House, where CMP is now located. The first Ford pumper to arrive in town was housed at this location increasing the town's ability to more effectively fight fires (This information is taken from Richard P. Mallett's History of Farmington – *Farmington Chronicle.)*

In 1912, A. B. Carr was Foreman & Chief Engineer, and A. D. Keith was First Assistant Foreman. The Village Corporation and the Town of Farmington shared the cost of fire department equipment; the Village Corporation supported the fire department through taxation. Should a fire occur outside the Village Corporation, or suburban limits which included West Farmington, the town paid for the firemen. In 1916 the fire department appropriated funds to acquire a Model T Ford chemical fire truck to enhance its firefighting capability.

In 1925, the fire department became Farmington Fire Company # 1. In 1931, a Maxim pumper truck, 500 GPM (gallons per minute) was purchased. On January 12, 1941, the fire department moved across the street to the former site of the Clifford Belcher home on High Street which was razed to build a more modern fire house where the Farmington Water Department is now located. At the annual town meeting on March 10, 1960, the Town of Farmington acquired the assets and liabilities of Farmington Fire Company #1 from the Farmington Village Corporation. From 1960 on, the fire department was known as the Farmington Fire Department. On February 11, 1975, the fire department voted to become the duly organized municipal fire department of Farmington, and on March 10, 1975 the town voted to accept responsibility for the fire department including disbursement of all funds and appropriations. Thus, the fire department officially became known as the Town of Farmington Municipal Fire Department. A year later in March 1976, the Farmington Falls Fire Department was accepted by the Town of Farmington as a sub-station of the Farmington Fire Department, and the town incorporated all land and buildings into the Farmington Fire Department. On December 12, 1976, the new fire station and municipal building opened on the former Alonzo P. Richards property on Lower Main Street. In 2000, the fire department became known as Farmington Fire Rescue with its first full-time Fire Chief.

Photo of 1916 Model "T" Chemical Truck
Courtesy of the Owls Head Transpotation Museum Owls Head, ME

Photo of "Engine 1" 1931 Maxim Pumper
Courtesy of the Farmington Fire Department

1860 Farmington's First Fire Engine "Eagle No. 1"
Rebuilt 1836 Hunneman fire engine with FARMINGTON painted
on the rear tool box in gold lettering
Photo Courtesy of the Henry Ford Museum Dearborn, Michigan

Before the Great Fire of '86

Looking up Main Street before the fire of 1886. Photo taken from present First National Bank corner. The J. J. Newberry Store now occupies a large part of the area shown. (Photo courtesy of The Knowlton & McLeary Co.)

Same scene as above taken in February 1969.

Before the Great Fire of '86

Looking down Broadway, Farmington before 1886. (Courtesy of Ben Butler)

Same scene as above taken in February 1969.

Scenes following Fires of '75 and '86

The Stoddard House after the Fire of 1875.   (Courtesy of Ben Butler)

Ruins of the Fire of 1886. Looking up Farmington's Main Street from F.S.C. Men's Dormitory.   (From collection of Mrs. Elizabeth Starbird of Strong)

3 CENTS PER COPY.

Wilton Record.

EXTRA.

VOL. VI.   WILTON, MAINE, OCTOBER 27, 1886.   NO. 274.

Outline Map, Showing the Location of the Ruins.

Cragin house.	Old Lake House.	
	Smith & Hunter.	
B. Sterry.	Worthley's.	
	Hotel Marble.	
E. Gerry	Common.	
	Unburnt District	
W. Tarbox.	included the build-	
	ings of T. H. Ad-	
Mrs. Hunter's.	ams & W.Tarbox	
	and one dw. at S.	
J. S. Milliken.	W. cor of Com-	
	mon.	
Varney.	Old Town House.	
	Lincoln & Rich'ds	
Drummonds.	H. L. Emery.	
	W. F. Belcher.	
John Lebash.	D. H. Knowlton.	
	K. M. & Co.	
Mrs. Morton's.	J. Gay.	
	Peo. Trust Co.	
Stovell's stable.	**BROADWAY**	BROADWAY.
	Stovell Stand.	
	Post-Office.	
	new Perkins blk.	
	Keyes blk.	
	Arcade blk.	
	old Perkins blk.	
	O. county buildg.	
	Russell & Priest.	
	McKeen.	
	Tarbox N.Y.Store	
	D. Hoyt.	
	A. D. Horn.	
	Bakery.	
	I. Russell.	
	McCleary house.	
	J. Matthieu.	
	Exchange Hotel.	
	Cong'l Parsonage	Baptist church.
	D. Pratt.	
	A. Belcher.	
	Knowlton house.	
	S. Belcher.	
	Cong'l church.	Methodist church
	S. Beedy.	John Allen's dws.
	old Swett dw.	Mrs. Johnson.
	F. G. Butler.	Mrs. Graves.
	D. H. Chandler.	
	Abel Russell.	

JAIL — PLEASANT STREET — Looking South — FRONT STREET

MAIN STREET — LOOKING SOUTH

The Unburned District within the radius of the fire is represented by the white space, excepting where otherwise stated.

October 27, 1886
Wilton Record.
Location of Ruins, Farmington's Great Fire
See Appendix for Full Article

Chapter 2

ESTABLISHMENT AND GROWTH
OF THE FARMINGTON FIRE DEPARTMENT

The first record of any serious fire in the Town of Farmington appears in Butler's "History of Farmington". It occurred in 1850 when the wooden buildings on the square between Exchange Street and Broadway were destroyed. The Stoyell Block located on the corner of Broadway and Main Street escaped this fire which was fortunate considering there was no organized fire protection. After this fire, a few individuals began demanding protection from the ravages of fire, but nothing came of it as the community was unwilling to tax itself for fire protection. It was June 1859 when the Farmington Village Corporation began taking measures to establish fire protection. The *Farmington Patriot* noted that an act to incorporate was accepted for the Farmington Village Corporation establishing regulations for the fire and police departments, but no action was taken to procure fire equipment. On August 7, 1859, another serious fire took place in a store owned by Francis Knowlton and occupied by True G. Whittier; this is currently the Knowlton McLeary Block. The fire spread in all directions, even though it was a seasonally calm night. The fire did $18,000.00 worth of damage and affected approximately 20 people and businesses. ($18,000.00 would be about $375,000.00 today.) Once again, concerned citizens rallied around the need for fire protection and went so far as to request the town purchase an engine and five hooks and ladders, but nothing happened. On December 29, 1859, a fire broke out on the west side of Main Street to a brick building with a tin roof that luckily was easier to contain and did not spread to other buildings. After the fires of 1859 the Village Corporation commissioned A.B. Caswell, Esq. to purchase a fire engine in 1860 at a cost of $400. (Some receipts show the cost at $600 which in either case would be about $10,000 today.)

On November 21st, 1860, Farmington's first fire engine arrived in town amid much publicity and fanfare. It was a rebuilt 1836 Hunneman Company hand pump fire engine previously owned by the Boston (Roxbury) MA fire department. In Boston, it had been known as "Salamader No. 5", but to Farmington it was "Eagle No. 1." Farmington now had its first fire engine, and an engine company was organized to work it. From the *Farmington Chronicle* dated Thursday, November 29, 1860, I quote:

> "Fire! By train of Friday last came the much talked of fire engine for the use of Farmington Center Village, and on Saturday afternoon the fire company, under the direction of "Captain Jennings", took possession of "der machine" at the depot, and two abreast escorted it to the village in fine style. Passing up Main Street, the "boys" greeted the officers of the corporation, assembled on the line of the march, with three rousing cheers, and passed on to the court house reservoir, where the hose was run off, the brakes manned, and

a trial of "Eagle #1", exhibited to the village audience, free of charge, and pronounced a "success". It is every way a trim piece of machinery, and will do good execution, if occasion requires, against its fiery enemy. The "Eagle's" members upward to forty as good fellows as the next place can scare up. Long live the Eagle Engine Company."

Again, I quote from the *Franklin Patriot* dated November 30, 1860 Volume III No. 45.

"A most beautiful fire engine, called the Eagle #1, paraded to our streets for the first time, on Saturday last, and the "boys" not only looked pleased, but acted like men with it. The engine drew water from a well twenty-six feet deep, and against a very strong wind with one hundred feet of hose, threw a stream of water through a seven and seven eighths inch nozzle, high above the spire above the Court House. Everyone seemed pleased with the first trial, and there can be no doubt that an efficient company has been organized, and a suitable machine furnished, by means of the Village Corporation. Hirrajh for the Eagle #1, and a big tiger for the first fire company of Farmington."

The "Eagle" was a horse-drawn, hand-pumped fire engine, dark red outside, with the word "Eagle" on each side. Equipment included a suction pipe, eduction pipe, a bell, metal holders for two buckets on each side and holders for three leather covered nozzles. Off the top of the box on each side were two suction holders and two suction pipes on the right. Supports for the suction pipes were curved metal rods connecting to the front axle. The rear brass tool box was marked in gold lettering with the name: FARMINGTON. The red 12-spoke wheels had black, gold and light red trim with brass hubs, a front wheel diameter of 31.25" and rear wheel diameter of 37.5". When a larger engine was needed, Farmington sold the "Eagle" to Atwood Box Company of Rockland, MA for fire protection. The Hunneman No. 169 "Eagle" is still around and part of a collection at the Henry Ford Museum, Dearborn, Michigan.

From the 1860's until the Great Fire of 1886, the fire department had only a small hand-pumped fire engine and a hook-and-ladder company. The engine house was erected on Academy Street, where the UMF Administration is now, before it was moved to Pleasant Street near the intersection of Front Street, Pleasant Street and Broadway. Reservoirs were dug in different sections of town to keep a supply of water for emergencies.

An interesting note is that in March of 1873 Captain F. V. Stewart of the Engine Company presented a paper to the citizens of Farmington on behalf of young men in the village who wanted to organize a brass band. One might assume some of those young men might have been associated with the Engine Company. On March 27, 1873, Captain Stewart sent out an appeal to the community for money to buy instruments. On April 24, 1873, E. I. Merrill was now Captain and he called a meeting at the Engine Hall with Hon. F. G. Butler presiding, and it was voted to fund money for the purchase of instruments. In June of 1873, the Coronet Band was formed which would later become Wheeler's Band. [The *Farmington Chronicle*] This is the first time a record of Captain F. V. Stewart's service to the fire company has been sourced between that of Captain Jennings in 1860 and Captain Merrill's in 1873. (Thank you to Nancy Porter for providing this information.)

On December 16, 1874, another destructive fire started on Main Street in a store owned by William Tarbox. Several buildings including a harness shop were destroyed. In 1875 Main Street was the site of yet another fire that started in a drug store and advanced to Broadway.

Then there was the Great Fire of 1886. From the *Wilton Record*, VOL. VI. NO. 274, dated October 27, 1886, I quote:

> "$300,000.00 FIRE AT FARMINGTON! - A Large Portion of our Sister Town Gone. - Including Three Churches, Post Office, Three Hotels, Forty-Two Business Concerns, and Thirty-Two Dwellings and Stables Connected. - SAD HAVOC ON EVERY HAND - Ninety-Six Families Homeless. Both Newspaper Establishments in the Same Condition the Record Office was in last February.

> Last Friday afternoon at about 3 o'clock, flames were simultaneously discovered by the construction crew of the railroad and by Captain E.I. Merrill's son Dana, bursting from a barn owned by the Stoyell heirs, situated on the west side of Front Street, near the narrow gauge railroad tracks. An alarm was immediately given, and in a few minutes nearly all the population was at work trying to extinguish the fire. The village has only a small hand fire engine and a hook and ladder company, both of which did good execution, considering the scarcity of water at hand, and at 4 o'clock the fire at the barn seemed to be entirely under control, although a strong guard was placed around it. At this time, it was generally thought that the danger was over, but it was not so ordained. The barn contained 45 tons of fine hay, farming tools, etc., and the sparks from the fire were carried in all directions, igniting the shingles of the adjoining wooden buildings, after having been fanned by the zephyrs that at times held high carnival. And at about 8:30 o'clock as Albert Sterry was standing in his yard he saw a light in E. Gerry's stable, and on inspection the building was all ablaze inside. Again, the alarm sounded and caught up by every alarm in town, and in a second all people were on a "wire edge". No effort was spared to control the fiery monster; but the efforts were of little affect for it spread like wild fire in every direction, and all the rest of the night things were sad to behold. Buildings were burning and falling on every hand, and men, women and children were mad with excitement, which is not surprising to a

party who has never experienced a burn-out. Furniture and valuable household goods were scattered to the winds, and unmerciless thieves seemed like ravenous wolves seeking to destroy what the fire should leave behind. It is said that thousands of dollars' worth of valuables was stolen. It does seem awful that at such times the land sharks will stoop to such measures as to pilfer from the penniless and homeless..."

At 9 o'clock Phillips was telephoned for aid, and at 10:30 o'clock Phillips' hand engine and hook and ladder company and 300 citizens from that place were in Farmington, having been drawn 18 miles in just one hour over the little railroad. We are told that they did considerable execution with their fire department. And no doubt Farmington is very grateful to them.

Lewiston and Portland were telephoned, for aid, and generously responded by immediately sending fire engines, and it is due to the good execution of them that more sets of buildings were not burned. Portland and Lewiston will never be forgotten by Farmingtonians. The county jail was consumed, but by remarkable forethought the turnkey, H. Jewell, freed the incarcerated prisoners and thus saved them from a terrible death. The old express horse and a cow that were in E. Gerry's stable were burned. Those were the only animate things that got singed that we know of, and it seems truly miraculous that it is so. It is said that the entire village seemed to be in a rain of fire, as if the fire was coming down from the clouds. Almost every house and store on Main St. were stripped of their contents and great damage was the result, where no fire trespassed. Farmington never saw such a season and it certainly does not wish to do so again. Another like conflagration would wipe the village from existence. Sad are the hearts of all her resident's now..."

[See the full *Wilton Record* article on the Great Fire of 1886 located in the Appendix. There is a detailed description of all the buildings, and their owners and tenants, that were affected. It is well worth a read. Of note, when the village began the rebuilding process, it was decided that buildings would be brick structures in the hope that brick buildings would be safer. FYI, $300,000.00 in 1886 would be approximately $6,000,000.00 today.]

At a special town meeting in November after the Great Fire of 1886, the town voted to appropriate a sufficient sum of money to purchase a steam engine and appliances for its use; a steam engine and equipment were purchased for $5,000.00 which today would be about $135,000.00. In December of 1886, the town tried to raise money to buy a fire pump and appliances to be operated by power generated from Charles H. Watson's water wheel which he agreed to furnish for free. It is not known if the money was raised for the fire pump, but town records indicate the selectman voted to make application to the state legislature for a reduction in state tax for the next two years for some town appropriations.

On November 26, 1886, George C. Purington became Foreman (Fire Chief) and I. Warren Merrill, First Assistant Foreman of the Farmington Steam Fire Company, No. 1. Other members of the Farmington Steam Fire Company No. 1 were: W. B. Elwell, 2nd Asst. Foreman; E. E. Jennings, 3rd Asst. Foreman; Chas. E. Wheeler, 1st Engineer; Chas. W. Marston, 2nd Engineer; Ruben Huse, First Fireman; C. A. Priest, 2nd Fireman; Arthur M. Bailey, First Pipe Man; Frank A. Davis, First Pipe Man; L.M. Burbank, Suction Hose Man; F. P. Adams, Suction Hose Man; and Firemen W. B. Bailey, Samuel Bailey, E. W. Bragg, Hannibal Berry, J. A. Blake, Fred A. Collins, F. E. Doughton, J.M. Matthieu, Carl Merrill, Dana Merrill, Joseph Pero, Geo. McL. Presson, Joseph Pero, John Robinson, and George E. Stevens.

The Farmington Steam Engine Company No. 1 had a steam fire engine, ladder wagon and hose carts. The Fire Engine Company was moved from the intersection of Front, Pleasant and Broadway to a building adjacent to the old Steam Laundry and the former showroom of J. W. & W. D. Barker, Inc. along High Street between Broadway and Church Streets. (See maps on following pages with engine house locations in 1885 and 1892.) Between 1887 and 1890 the horse-drawn steamer fire engine arrived in Farmington along with a ladder wagon and hose carts; the exact date is not known. The horses were so well trained that when an alarm came in, the horses were unhitched and collared, and they would automatically take their place on the tongue or pole of the apparatus. The harnesses were hung by ropes and pulleys above the wagon tongue and were lowered over the horses' backs and were hitched in place and ready to go in a very few minutes. Remember, that was over 100 years ago, when there were horses and horse stables all over town. I have found no written accounts of fire fighting in this era, but it sure would have been interesting to read. It should be noted that many of the early records of the Farmington Village Corporation were either lost or destroyed.

**FARMINGTON FIRE DEPARTMENT
EARLY 1900'S**

Farmington Steam Engine Co., No. 1 circa late 1800's Seated Everett Clark Standing L-R: Carl Merrill, Frank Davis, Harry Knapp, Bert Spinney, E.M. Higgins, Rufus Jennings, E.A. Odell, J. M. Small, Ben Hayes, Robert Campbell, Geo. Dobbins. Picture Courtesy of Town Annual Report

Map Courtesy of Nancy Porter

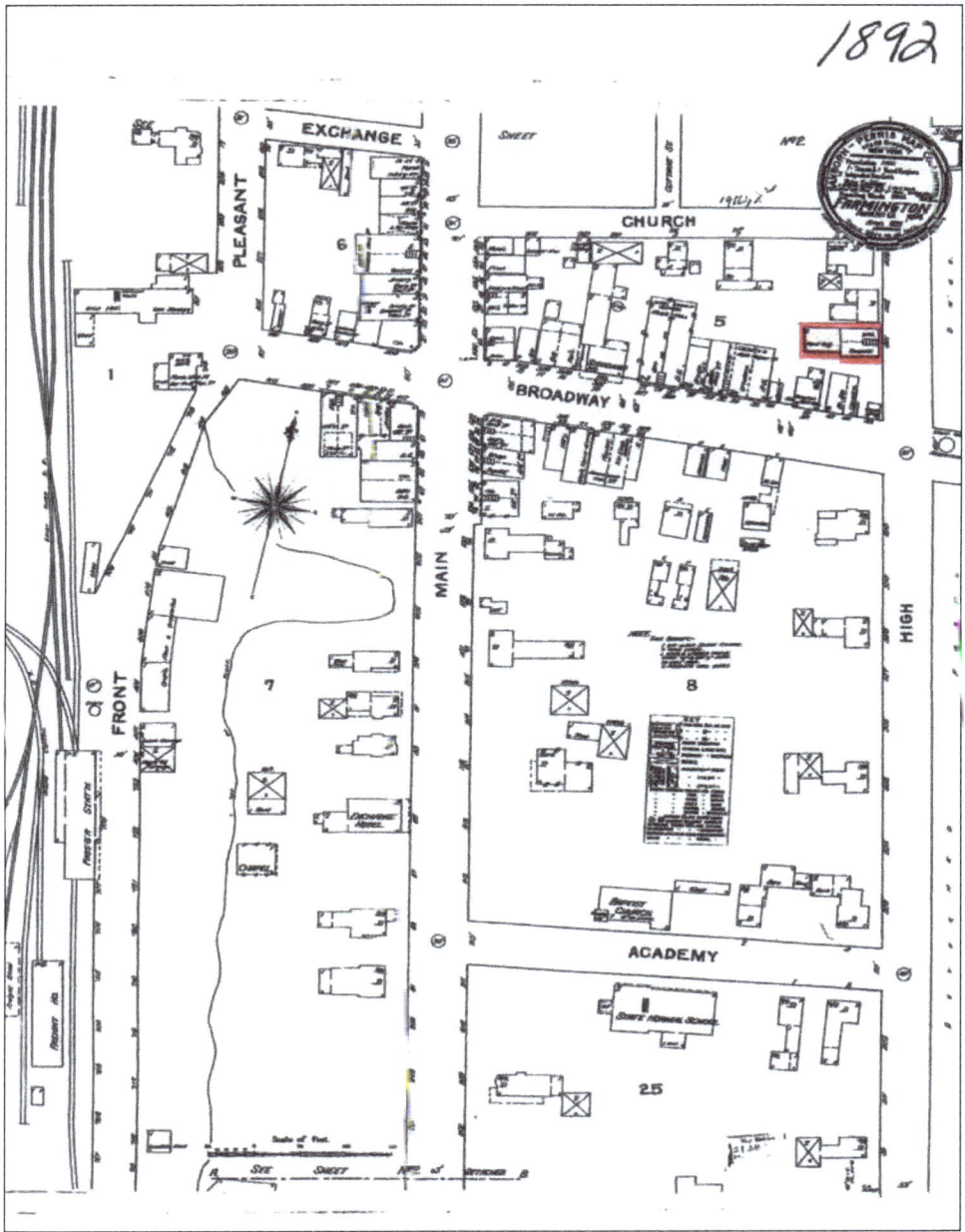

1892

Map Courtesy of Nancy Porter

Farmington Fire Company at Station near corner of Broadway and west side of High Street, circa 1906.

Photo Courtesy of Town of Farmington

Chapter 3
FARMINGTON FIRE DEPARTMENT EVENTS AND EQUIPMENT

I will list as near as possible some of the events that have occurred in the Fire Department over the past hundred years. I will also list the addition of auxiliary equipment.

The hydrant system in Farmington was started about 1905 after two (2) eight inch water lines were laid from Varnum Pond. Hydrants were placed along different streets, and at street corners and major intersections. Ten-inch water lines ran from the hydrants and were then connected to 8-inch water lines to the reservoir. In 1907, 10 hydrants had been installed in West Farmington.

In 1906 the town appropriated $200 for the fire department and $300 for the fire company; George H. Blake was the Foreman and Chief Engineer. The fire alarm system was changed from a series to a bridging system. It was noted in the minutes of the water department from December 31, 1906 that: "The fire protection of Farmington Village probably is not equaled in the state. A reservoir on either side of us, a steam fire engine, pumping plant, gravity system, and fire ladders ready at a moment's warning to fight any fire to a finish, should tend to make our insurance rates lower."

From 1900 to 1915, I have little knowledge of the changes in equipment from horse drawn to motorized fire equipment. However, we know from Natalie and Benjamin Butler's "History of Farmington", during 1912, the first Ford pumper was housed with the steam fire engine in a new fire station built by the Village Corporation on the corner of High and Perham Streets, across from the Octagon House, where CMP is now located.

In 1912 an audible air horn was installed on the fire station (formally the CMP building on High Street) as a result of alarm bell malfunctions which had caused delays in calling out the fire department to fires. The fire department began investigating a new alert system after the 1900 A. H. Abbott fire that consumed two buildings (Abbott Library and Little Blue School for Boys), and surrounding shrubs, vines and trees.

In 1914, the town appropriated $200 for the fire department and $500 for the fire company. In 1919 the town increased the fire department appropriation to $350 and the fire company remained at $500.

In 1916, a Model T Ford chemical fire truck was added to the department, and today can be found on display in the Owls Head Transportation Museum, Owls Head, Maine. The Model T Ford fire truck was a popular choice for fire departments in small towns with modest budgets.

In the 1920's a hose-and-ladder truck (Dodge) was added.

On January 1, 1923, Victor C. Huart was elected foreman and chief engineer (Fire Chief).

On October 19, 1931, a new Maxim Pumper 500 GPM (gallons per minute) fire truck was purchased, and the Dodge ladder truck was removed from service and stored in Carroll Collins barn on Perham Street. The new Maxim Pumper was the top of the line fire truck at the time, is still operational, and is a great show piece. On November 15, 1931, the Maxim crew held a one hour practice session near the Common, pumping water from a well, opposite the courthouse, which is now covered by Main Street, and then pumping water from a hydrant. This may be the first time the new pumper was used.

On March 1, 1932, the fire company talked about the need for an insurance and liability policy.

On May 4, 1937, a need was discussed for a new Dodge fire truck that would carry chemicals to greatly assist on-site fire fighting capabilities.

In 1939, the fire department payroll was $600.00 to be divided among the chief, janitor and clerk/treasurer.

J. Bauer Small was accepted into membership in 1939. He would become Fire Chief in 1951 and serve as chief for 18 years.

In 1940, the Maine Department of Education sent a letter urging all firefighters to attend a training session at Camp Keyes in Augusta from June 21st – 23rd. On September 3rd, talks ensued about using "gas masks" at fires and developing training on how to use them. In October, house uniforms were offered by an interested person and questions of who would pay for them going forward were discussed.

On December 3, 1940, Farmington Fire Co. No. 1 began the move to the new fire station on High Street.

On January 12, 1941, a new, more up-to-date, fire station opened on High Street where the Farmington Water Department is located; the public was invited to an Open House. On June 18th and 25th, training drills were conducted on the use of ladders, pumping, and hoses. On November 4, 1941, the new "fog nozzle" was demonstrated, and later purchased for the department. The new fog nozzle was said to be more effective because the water dispersed formed smaller droplets which absorbed more heat. However, there were some reports of increased steam burns to firefighters.

On January 6, 1942, there were discussions about acquiring an "Auxiliary Truck", but no action was taken. In October 1942, the Maine Fire Chief's Association held their meeting in Farmington with more than 100 in attendance. This was the FIRST for our Town.

In August 1944, the company still has 2 trucks, the Maxim Pumper and Dodge hose-and-ladder.

During the WWII years (1939-1945), many members of the fire department were called into service, and upon their return from military service they were re-instated into the fire department. Firemen who remained at home and manned the fire department conducted scrap drives to help the war effort.

Victor C. Huart

Above Photo L-R: Del Johnson, Chief Huart, J. Bauer Small, Uknown. Circa 1940-1950's
Photos Courtesy Farmington FD

Above Photo Fire and Fireman unknown

Above Photo Fire Chief Bauer Small, white chief helmet, others unknown

VENING JOURNAL   TUESDAY, NOVEMBER 16, 1943
TIN COLLECTED BY FIREMEN AT FARMINGTON, LAST SUNDAY.

World War II Firemen Scrap Drive Artic e Courtesy of Don DeRoche & Franklin Journal

Victor C. Huart
Farmington's Longest Serving Fire Chief
January 1923-March 1951

On February 5, 1946 membership for the department was raised to 25 men.

As of May 5, 1947, the old roll call was eliminated, except for out-of-town calls, due to having 2 trucks answering outside calls.

In 1948, a Chevrolet 500 gallon water tanker was purchased for added protection outside the hydrant district. This now made for 3 trucks in the department: Maxim Pumper, Dodge hose-and-ladder, and 500 gallon tanker. The water tanker was added to the fleet as a result of the wild fires that spread throughout Maine in 1947 burning over 200,000 acres of forested land and destroying 9 towns and 1,100 homes along the coast as far north as Bar Harbor and as far south as Kennebunkport. Many small Maine towns were ill-prepared to fight a fire of this magnitude as they did not have a fire department, fire equipment, mutual aid agreements with neighboring towns, or adequate training.

Scott Air Pak 1800 psi with 15-min air tank. First two arrived at Station in 1951

On March 27, 1951, Fire Chief Victor C. Huart died. Chief Huart was elected fire chief in 1923, and held that position for 28 years. Chief Huart remains Farmington's longest acting fire chief. His final "ride" was on the old Dodge truck as he wished. After the burial ceremony, the company returned to the station and gave two blasts of the horn; the "ALL OUT" was sounded.

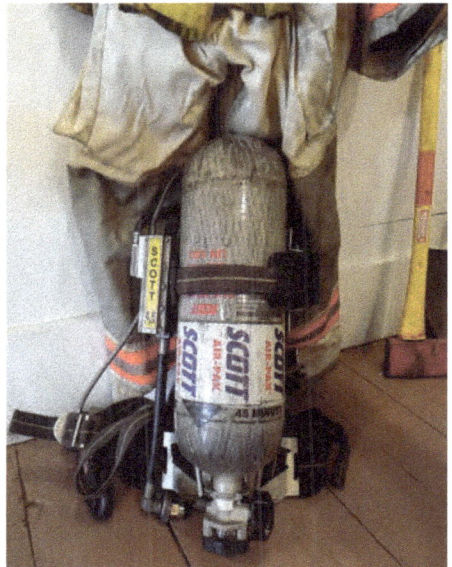

Newer model Scott Air Pak 4500 psi with 30 - 45-min air tank

On April 2, 1951 at a special meeting J. Ambrose Compton was elected the new chief, but 15 minutes later he resigned; his was the shortest tenure. Duane Hardy was secretary. Shortly thereafter, J. Bauer Small was elected Fire Chief, and served as chief for 18 years from 1951 to 1969 with a total of 29 years of service.

J. Bauer Small Fire Chief April 1951 - December 1969

In July 1951, the first two 1800 psi Scott Air-Paks were put into service with 15-minute air tanks and anyone interested in learning to use them was invited to training sessions. Today our air tank capacity is 30 minutes or larger with 4500 psi.

In February 1954, it was voted to purchase a Ford F 750 fire truck from Farrar Co. of Massachusetts with a 500 GPM front-mount pump, and a tank carrying 750 gallons of water called the "water thief". The truck arrived in the late summer of 1955 and was called Engine # 3. A deck gun was also added to this truck and is now on the current Engine 5 Mack fire truck.

In 1954, Fire Chief J. Bauer Small submitted the first report of the fire department to be included in the annual town report. I quote Chief J. Bauer Small:

"This has been due to the fact the Fire Company is maintained by the Farmington Village Corporation. By maintenance, I mean they provide the station, equip us with boots, coats, etc., and pay us a set sum each year for all the fires inside the Corporation limits. As the town proper contributes its share toward the operation of the Department I feel it only right to inform them of our activities."

In 1955 the town appropriated $2000.00 a year for the fire department.

On June 3, 1958, new audible alert horns were installed on the High Street fire station.

On February 2, 1960, I was accepted as a Rookie and became a full member in 1961 although I was unofficially helping the department fight fires before this time. I had been unable to join the department prior to 1960 as I resided just outside the Village Corporation limits. For some humorous moments of my involvement with the fire department see Chapter 14.

At the March 10, 1960, annual town meeting, the Town of Farmington voted to acquire the assets and liabilities of the fire department from the Farmington Village Corporation. Until 1960 the Village Corporation had controlled the fire department.

In March 1961, Chief Small reported on the different call-type "quick alert" radio systems for firefighters to have in their homes; and in 1962, the fire department began using the house alerting Radio Paging System (Plectron) for calling firemen, gradually eliminating the audible horn. For more information on the town fire alarm and alert systems see Chapter 15.

April 5, 1961, Chief Bauer Small announced that the Town of Farmington was now responsible for paying all fire department bills.

In 1963, the Farmington Fire Department joined the newly organized Franklin County Firemen's Association. Farmington Chief Bauer Small was one of the organizers of this association as well as one of its first officers.

In August 1963, Radios were distributed to firemen, in groups of 5, starting at the top of the membership roster with radio rotation every 6 months to firemen down the list.

On May 22, 1964, Fireman Robert Marquis died suddenly. He had been a member of the department for 2 ½ years.

On October 25, 1965, Chief Small reported that the new Ford Thibault engine had left Tennessee and would be arriving soon. This would be a 750 GPM pumper and carry 700 gallons of water. This was Engine #4, located at the Farmington Falls station.

In 1968, a new respirator was purchased replacing the old outdated one.

In 1968, Chief J. Bauer Small's salary was $300.

On December 2, 1969, Chief J. Bauer Small submitted his resignation. At the annual meeting, Forest L. Allen, a member of the department since 1943, was elected the new Fire Chief where he served for 8 years.

In 1969, Mrs. Forest "Vertie" Allen began dispatching out of her house for the fire department. Vertie's was the clear voice that came over the radio for the calls. She continued as dispatcher until November 1982 when dispatch was taken over by the Sheriff's Department. After her retirement Vertie Allen continued to volunteer for the fire department giving out burn permits until 1991. In 1970, the West Farmington Fire Department merged with the Farmington Fire Department.

In 1971, the Company acquired a 60 foot Bangor Extension Ladder and foam nozzle. The Red Cross offered a first aid course for the firemen.

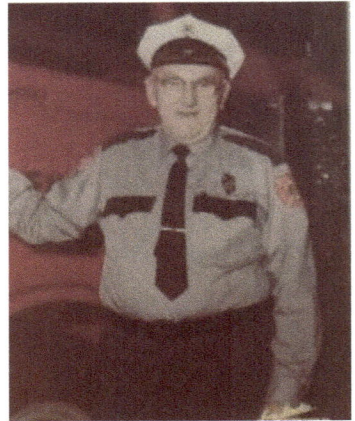

Forest L. Allen Fire Chief
December 1969 - April 1977

May 4, 1971, S. Clyde Ross was accepted as Rookie.

April 25, 1972, members of the Farmington Fire Department attended the funeral of one of its own, Duane Hardy, at the Old South Church.

On September 8-9th, Francis Roderick of Waterville taught the uses of ground ladders and the new Bangor Ladder. On October 30th, three more alert radios were purchased.

On November 7, 1972, a committee was formed to study the possibility of building a new fire station. Fire Department Building Committee Members were: Forest Allen, Harrison Bragdon, Harvey Smith, Richard Russell and Pete Durrell. Many hours and days of serious deliberation followed.

Vertie Allen Dispatcher 13 years
1969 - 1982

On April 3, 1973, an option to purchase was signed on the Alonzo P. Richards (former Maurice Taylor) property on Lower Main Street for a new Fire Station -Municipal Building complex. See town Study Committee report in the Appendix.

In December 1973, the fire department made a request of the town for a new fire truck.

On March 19, 1974 requirements for a new Mack truck were discussed which would become the Mack Engine #5. The cost of this truck would be $53,250.

In 1974, training was initiated for resuscitation and air packs, and changes were adopted for safety boots and turn-out gear for the firemen. A new power saw was also purchased.

At the August 6, 1974 Special Town Meeting, the voters approved the new fire station - municipal building complex for the A.P. Alonzo property on Lower Main Street.

During late 1974 and early 1975, the Insurance Service Organization conducted a survey of the fire department and local water supply. After the evaluation was completed, Farmington received a new and improved rating of C-5, which benefited many property owners with lower insurance premiums. New equipment, more training, and better facilities were factors in determining the town's fire rating. In 1974, Insurance Underwriters viewed equipment 20 years of age and older obsolete. Today equipment is considered obsolete after 15 years so it's important to maintain adequate reserve funds for new equipment in each year's budget.

On February 11, 1975, the fire department voted to become the duly organized Municipal Fire Department of Farmington, and on March 10, 1975 the town voted to accept responsibility for the fire department including disbursement of all funds and appropriations. Thus, the fire department officially became known as the Town of Farmington Municipal Fire Department.

May 20, 1975, Town Manager, Philip Schenck explained new fire department rules and they were adopted.

June 3, 1975, the members of the Farmington Fire Department attended the funeral of former Assistant Chief Maurice Taylor.

At the meeting on June 3, 1975, all firemen present were duly sworn in as firefighters for the Town of Farmington by Town Clerk, Fay Adams, establishing their authority to act on behalf of Farmington.

On June 16, 1975, the new fire station and municipal building were accepted by the town with 118 vote margin; the cost of the new building would be $807,380.00.

In 1975 training sessions were held in relay pumping, draft pumping, and the use of new equipment.

On December 6, 1975, the new Mack Pumper 1000 GPM arrived to be known as the new Engine #5.

On February 3, 1976, all members of the fire department received dress uniforms which were to be purchased from Allen Uniform Company in South Portland, ME.

On March 2, 1976 by a vote at the annual town meeting, the Farmington Falls Fire Station was accepted by the Town of Farmington as a sub-station thereby incorporating all land and buildings into the Fire Department.

In the summer of 1976, the firemen dedicated themselves to raising money for the Pine Tree Burn Treatment Center by participating in walk-a-thons, holding chicken barbecues, manning the Fire Department Fair Booth, etc. New rules and regulations were accepted by the department.

1976 Franklin County Firemen's Assn. gift to CMMC Pine Tree Burn Treatment Center. Pictured L-R Front Row: Carroll Corbin, Madeline Massey of CMMC. Back Row L-R: Sheridan Hargraves, Mrs. Carroll Corbin, Forest Allen, Morrill Collins, Harrison Bragdon, Robert McCleery, Lewis Holbrook, Clinton Blaisdell

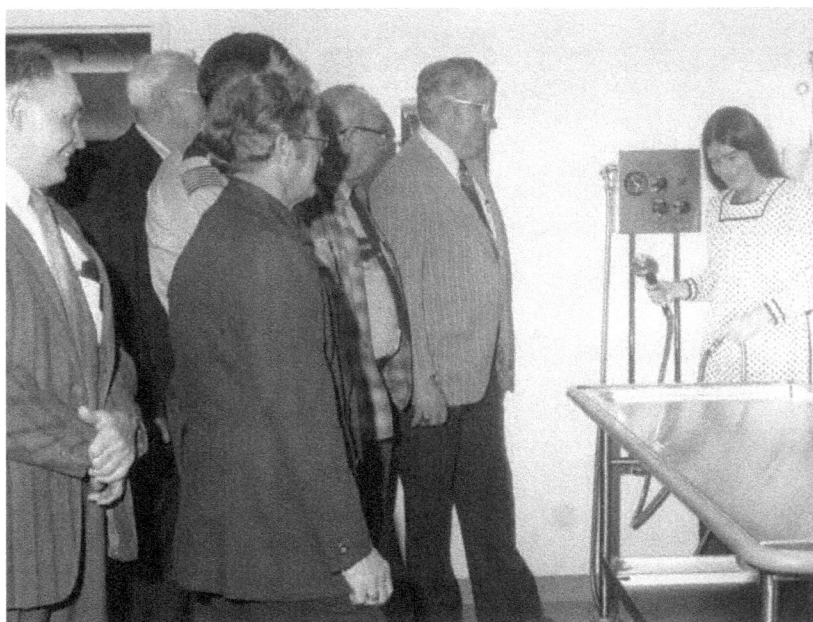

CMMC Madeline Massey demonstrates for Franklin County Firemen's delegation the debridement table & equipment for the treatment of 1st and 2nd Degree Burns.

Harrison Bragdon
Assistant Chief
1968 - 1976
FFD 1st Training Officer

On December 12, 1976, all apparatus and equipment were moved from the High Street Station to the new station on Lower Main Street. The next day, December 13, 1976, Assistant Fire Chief Harrison Bragdon died suddenly of a heart attack on his way home to his farm in Industry. Assistant Fire Chief Bragdon was a member of the fire department for 25 years and had served on many committees as well as being the first Training Officer for the Farmington Fire Department.

In 1976, fire department salaries totaled $550.

On January 23, 1977, an Open House was conducted for the public at the new fire station and municipal building on Lower Main Street.

March 3, 1977, Glenwood Farmer was promoted to the rank of Assistant Fire Chief (Deputy Chief) for Operations and Training replacing the late Harrison Bragdon.

On April 1, 1977, I was appointed Fire Chief succeeding Chief Forest L. Allen who had retired after 8 years as chief with 34 years of service. I had entered the department as a Rookie in 1960, was elected Secretary Treasurer in 1974, promoted to Assistant Chief of Administration and Finance on April 1, 1976, and in 1977 served as Franklin County vice president of the Maine State Federation of Firefighters.

Between April 5 and July 5, 1977, Deputy Chief Glenwood "Glen" Farmer was appointed Training Officer, and Richard Russell, Assistant Chief of Administration and Finance.

On September 6, 1977, the new Squad Truck was fully equipped and the crew ready to go. October 12th saw another Public Open House at the fire station during Fire Prevention Week.

On November 2nd, Dr. Paul Brinkman conducted physical examinations of the firemen.

By the end of 1977, all town fire alarms had been tested and were fully operational.

In 1977, I began restructuring the fire department into a chain of command system going from what had previously been a Chief and two Assistant Chiefs to the ranks of Deputy Chief, Captain, Lieutenant and Private. I believed this change would enhance morale and retention by offering more opportunity for promotion and advancement of firemen.

On April 5, 1977, Terry S. Bell, Jr. was accepted as a Rookie into the fire department. He became a Lieutenant in 1983, then promoted to Assistant Chief in 1984, Deputy Chief in 1991, and my replacement and the town's first full-time fire chief in 2000.

In 1977, fire department salaries were $575 and increased to $630 in 1978.

On February 6, 1978, the fire department switched over to the State Fire Marshal's Formal State Fire Reporting System which used a standardized incident reporting system versus our system of reporting fires in our record books. This statewide system provided local chiefs such as myself with help in determining fire trends, and specific areas of fire loss prevention. Training on the new system was provided.

February 25, 1978 a fire truck accident occurred responding to a call at the foot of the Hill Street rotary near the bridge that sent three firemen to Franklin Memorial Hospital. The three firemen were: Jack Bell, S. Clyde Ross and Terry Warren. No other vehicles were involved.

In February 1978, S. Clyde Ross was appointed Assistant Chief for Public Relations and Fire Prevention. Other captains and lieutenants were also named. In May, the town gave permission to add six more firemen to the roster, increasing the number of firemen to 45, if needed. In 1978, more attention was given to extensive training programs.

On March 1, 1978, Tim Hardy was accepted as a member of the fire department. He was promoted to Lieutenant in 1984, Assistant Chief in 1991, Deputy Chief in 2000, and now serves as the Director of the Franklin County Emergency Management Agency.

On April 14, 1978, Dick Cadwell, State Fire Training Service, conducted a training session on Front Street for our firemen. Temple and Wilton also participated in the exercise and Industry observed. This was simulated major disaster training. As the training ended, a fire call came in for the Gilbert property on the Back Falls Rd.

On December 5, 1978, Steve Bunker joined the fire department.

In January 1979 an editorial appeared in the Franklin Journal praising the professionalism of our entire organization from dispatch to response. This was just one of many the fire department received over the years.

On a cold, windy February night in 1979 two calls came in at the same time: one a radio call for a mobile home fire on Titcomb Hill and the other a call for a fire on Perkins Street. After the fires were out, much later in the evening, the Farmington Falls firemen served an oyster stew supper for the firemen which was much appreciated.

On October 9, 1979, a fire prevention program, to be used with local elementary school children, was instituted and graciously supported by local insurance agencies. "The Captain No-Burn Program" was well received, and more than 300 students participated. Assistant Chief S. Clyde Ross served as Public Relations interface to the schools; in 1985, there were 580 elementary school children involved in the program, and by 1996 1,200 school children participated in grades K-8.

In October 1979, the fire station once again opened its doors to the public during Fire Prevention Week.

In 1979, fire department salaries were $3,534.

In 1979, a new aerial ladder truck arrived on December 20th; and the fire department purchased its first air compressor from Ernie Goslin to fill air bottles.

In 1980, I was elected president of the Franklin County Firemen's Association and helped organize and facilitate the Franklin County Attack School and Mutual Aid agreements throughout the county. Farmington was the first town to offer Firefighter I and II State Fire Academy training for adults on weekends and evenings rather than having firefighters attend 2 weeks of training in August at Presque Isle. The fire department also implemented and supported fire training in the use of Self-Contained Breathing Apparatus (S.C.B.A.).

On May 10, 1980, Wilton Academy was destroyed by fire. The Mutual Aid call went out to East Wilton, East Dixfield, Jay, and Farmington with over 100 volunteer firemen responding. The Farmington Fire Department appeared on scene with the Squad Truck, Engines 4 & 5, and the new 75 foot aerial ladder truck. Thirty-one (31) of Farmington's firemen battled the blaze for 6 ½ hours and Fireman Kenneth Durrell climbed the 75 foot aerial ladder to put water on the roof. Assistant Chief Glenwood Farmer, several other firefighters, and myself returned into the night to watch over the smoldering building when the Wilton Fire Department was called to another fire. Six hundred and fifty (650) feet of 1 ½in hose, fifteen hundred (1,500) feet of 2 ½in hose, one thousand (1,000) feet of large diameter hose, and 25 bottles of air were used by Farmington firemen.

Structure Damaged Heavily By Fire

FARMINGTON — Facts from Fire Chief Robert L. McCleery Thursday verified the first report of late Wednesday evening that the story and a half wooden frame house that was extensively damaged by fire on the Back Falls Road was unoccupied. However, contrary to the first report received to the height of the blaze, the vacancy had been only since February.

McCleery said that Mr. and Mrs. Paul Gilbert and family had resided in the home until February when the family moved to Livermore Falls. The house is part of the property of Gilbert's late father, Frank Gilbert. A daughter, Mrs. Olive Taylor with her husband and children, reside next door in the home occupied by her father, Mr. Gilbert, until his death.

Mrs. Taylor apparently noticed a light in the vacant house and notified police about 10:30 p.m. that she could see a flickering light in the second floor window. The house is located nearly a quarter of a mile from U.S. Rte. 2 and State Rte. 4 from where hoses were laid from a fire hydrant near the main traveled highway.

McCleery stated that at least 20 firemen responded to the call, bringing the Squad Truck and four engines. Hawthorne Ambulance was also at the scene and the Central Maine Power Company dispatched a crew and truck to the area.

McCleery said it was discovered that the fire started at the foot of the stairs in an electrical box. The fire was drawn up the stair well and demolished the second floor, burning through the roof which was sagging by the time the fire was extinguished. He said very little fire damage was noted on the first floor of the home.

The firemen were just returning to the fire station from a training session with Dick Cadwell from the State Department Fire Training Service when the house was called in. The men had been working on a simulated major disaster reported on Front Street in the vicinity of the Farmington Farmers' Union Store. The exercise in which Wilton and Temple firemen also participated, was planned so that the firemen were hampered by various problems in covering the disaster to the best advantage. After each phase of the simulation the firemen would congregate to compare notes and discuss what they had done wrong. Firemen from Industry observed the training session.

McCleery said Thursday he was "well pleased" with the training session and also with the work of the firemen at the real fire which followed.

Farmington Police and State Police directed traffic on Rtes. 2 and 4 and kept the Back Falls Road clear of vehicles for the fire equipment. The hoses were strung across the busy highway holding up moving vehicles for a time.

The firemen had just about gotten back to the fire station when the car fire was called in.

The Lewiston Daily Sun

Lewiston, Maine   Friday, April 14, 1978   21

Photo & Article Courtesy of the Lewiston Daily Sun.

To the Editor:

This letter is one of appreciation and acknowledgement for the speedy response, competent action and the sincere concern of the Farmington Fire Department.

On Wednesday night, January 17th I was visiting in Farmington from Minnesota with my family, the Allen Flints. Mary and Allen were away for two days and I was babysitting with Joshua, age 7.

The Flints live in a house built in 1868. It is a two story home and built of wood. As you all know the wind was blowing hard and there was lots of snow. Many peculiar and rather frightening noises were coming up from the furnace room. It was late and the 11 o'clock news was over and the noises seemed to be increasing and some motor was going almost continuously and seemed to be getting louder and louder with each "whoosh" and I was scared.

The lights were out next door so I looked in the telephone book for the fire department to ask them for advice about the furnace. A most courteous lady answered the phone and then referred me to someone else. A gentleman listened to my story and then answered that they would send someone out to check into the situation. Two men came out very quickly. I asked them their names and they were Chief Robert McCleary and Deputy Chief Glen Farmer. They checked the furnace thoroughly and assured me that there was nothing wrong. They also gave me the name of the furnaceman to call in the morning if necessary. They were most businesslike and courteous and thorough.

That night I learned that Farmington had a volunteer fire department. Their help was as fast and as good if not better than any professional fire department with which I am familiar.

My congratulations to the Town of Farmington and to the Fire Department for having such an able organization. Thank you.

Sincerely,
Mrs. S. E. Halpern
St. Paul, Minn.

January 1979
Dispatcher Vertie Allen answered call, referred to retired Chief Forest Allen, who relayed to Chief Bob McCleery and Deputy Chief Glen Farmer.

Editorial Courtesy of the Franklin Journal

In May 1980, the Farmington Fire Department arson squad investigators were Captain Harold "Stub" Hemingway and myself. Clyde Barker, fire investigator for Wilton Fire Department who had just completed arson investigator training at the Police Academy assisted with the Wilton Academy investigation along with Bernard Emery, State Police Fire Marshal's office, Mac Burdin, state investigator, Blake McKay, state electrical inspector, Roy Lesson, Franklin County Sheriff's office, Wade Atwood, Wilton Fire Chief, and Earl Brown, Wilton Assistant Chief.

On August 26, 1980, a new fire truck committee was appointed to draw up a proposal for a new 1000 GPM pumper. On April 14, 1981, the town agreed to purchase a new truck by Mack Co. at a cost of $106,000.00.

In May 1981, the Farmington Firemen's Ladies Auxiliary became an official entity whose purpose it was to aid and support firemen with food and drinks during serious fires, and related calls, day or night. Over the years, the wives of firemen as well as other concerned citizens had come to the aid of firemen while fighting fires. Florence Norton King was a fixture at the fire house preparing soups, food, and coffee when fires occurred during the tenure of Vic Huart and Bauer Small. Sandy Knight, wife of fireman, Richard Knight, decided to formally organize the Ladies Auxiliary, and she had my full support and encouragement. This group not only provides coffee and food when needed, they also raise funds for fire department apparatus, donate to local charitable organization, and help with prizes and snacks for the children's annual Easter Egg Hunt.

Arson Investigators Captain Harold "Stub" Hemingway and Chief Bob McCleery

Flames Destroy Wilton Academy

Sentinel Correspondent

WILTON — Wilton Academy, a 151-year-old local landmark where generations of local residents went to school, was destroyed in a major fire Saturday.

Seven firemen from five area departments were treated for minor injuries and smoke inhalation at the blaze, first spotted at 2:14 p.m. by a newspaper correspondent.

The main building, which once was used as a church, was leveled in the fire. An addition where classrooms and a science laboratory were located was damaged but still standing Sunday.

THE WIND-SWEPT blaze quickly engulfed the structure at the junction of Depot Street and Old Route 2, despite the efforts of firemen from Wilton, East Wilton, East Dixfield, Jay and Farmington.

Wilton Fire Chief Wade Atwood, who took charge of the fire fight, said Sunday the cause of the fire is still under investigation.

It was unknown how long the fire had

Wilton Landmark
Mourned: Page 13

been burning, he said, when it was first spotted.

Mrs. Jean King of Wilton, who turned in the alarm and who lives near the academy, said two girls came to her door at about 2 p.m. Saturday to report there was smoke coming from the three-story building.

Mrs. King said she went to the academy, saw smoke coming from around the main entrance and called firefighters from a nearby home.

THE FIRST UNITS, from Wilton, were on the scene in minutes, but by then flames were already visible through the windows.

As the fire grew in intensity, Mrs. King said, windows began breaking out and flames appeared everywhere.

Ladder and tank trucks were pressed into service as more than 100 firefighters, the majority of them volunteers, battled the blaze.

Among those treated at Franklin Memorial Hospital in Farmington were Cary Pond and Gregg Oakes, both of Wilton, who suffered smoke inhalation. Oakes was admitted overnight and discharged Sunday.

FIVE OTHERS from the area were also treated at the scene for smoke inhalation and cuts and abrasions. There were no serious injuries.

Hundreds watched the spectacle, some of them weeping, as the fire raged out of control.

The academy, once a private school, had been last used as a middle school by School Administrative District 9. SAD 9 officials met later Sunday to determine where to send the displaced youngsters.

Firemen were still at the remains of the church-turned-school Sunday as lines of cars filled with onlookers moved past.

Chief Atwood estimated 300,000 gallons of water were used in the fire fight. He commended cooperating fire departments and emergency personnel who participated in battling the blaze.

Sentinel Photo by Jean King

The cupola atop Wilton Academy's main building totters into an inferno of flame despite efforts by more than 100 firemen to save the 151-year-old landmark. The fire razed the one-time church and damaged an addition containing classrooms.

Above: Courtesy of Waterville Morning Sentinel.

Left: Courtesy of Barbara Yeaton and Franklin Journal.

Robert L. McCleery 37

From 1990's to present [2000], the Firemen's Benevolent Association and the Ladies Auxiliary work together on the annual Easter Egg Hunt for boys and girls in town, 10 years old and younger. They place 100 eggs about town and provide clues to find them. This creates a lot of interest. Each egg contains a slip noting the prize won. There are also grand prizes, usually bicycles, presented to a couple of winners. Captain Jack Bell's plumbing company supplies bikes to the winners on behalf of the Benevolent Association.

In 1981, Assistant Chief S. Clyde Ross role as Public Relations and Fire Prevention liaison expanded to include the submission of a Training Report for the fire department to be included with the Report of the Fire Chief each year in the town's annual report. (See reports in Appendix.)

In July 1981, the fire department organized and hosted the First Franklin County Fire Attack School and invited state instructors to lead workshops along with our own instructors. The Fire Department continued to host through 1988 until local companies began providing their own training. In the year end annual report, I said of this event:

"Several of our men were instructors for this well-attended session. We gained much knowledge as well as recognition from other departments when they realized the expertise of our men and the equipment we have for fire attack. Farmington can be very proud of its men and equipment."

In August 1981 Terry Bell attended the Maine State Fire Academy in Presque Isle where he was certified as a Fire Fighter I and II. This was an honor, and first, for the department.

In 1981, several firemen attended the Fire Attack School at the University of Maine at Orono. In 1982, we added a second Mack truck, Engine 6, to the fleet with a 750 gallon water tank.

In February 1982, the Benevolent Association of the fire department purchased a new air compressor to compliment the "Cascade System" in the department's breathable air service operation. The department could now provide air, at a cost, to any user. This greatly increased Franklin County firefighter's use of S.C.B.A. (self-contained breathing apparatus) which safety and prudence require in immediately dangerous to life or health atmospheres.

On May 23, 1982 seventeen firemen attended the Henderson Memorial Baptist Church where they were recognized and honored for their service to the community. In 1982 Firemen Tim Hardy and Steve Bunker attended Fire Fighter I & II training in Presque Isle.

Girl Scout Troop 921 presented the Farmington Fire Department with a plaque acknowledging our firemen's service to the community.

On December 7, 1982 Forest and Vertie Allen were honored for their 40 years of service to the Town of Farmington. The Allens received the thanks and appreciation of the citizens of Farmington. The 1982 Farmington Town Report was dedicated to Mrs. Vertie Allen: "for her 13 years as Fire Department Dispatcher and for her faithful service to the citizens of Farmington in attending to their needs for emergency assistance."

Pictured L-R 1st Row: Cheryl Walker Bunker, Linda Brown, Sharon Barker, Eletrice Farmer.
2nd Row: Sandy Knight, Lois Bubier, Loretta Bard, Missy Richards, Rachel Fronk, Linda Chabot.
3rd Row: Melissa Allen, Glen Farmer.

1982 Girl Scout Troop 921 Honors Farmington Firemen Pictured L-R: Angela McCafferty, Sandra Meader, Rebecca Hunter, Chief McCleery, Michelle Pratt, Jill Casey

In April 1983, the Farmington Fire Department hosted a pilot program to train fire department personnel on Instructor Level I Training to become certified fire instructors in their own departments. Twenty firemen attended including six from Farmington. The Farmington firemen were: Steve Bunker, Richard Knight, Nelson Collins, Malhon Moore, S. Clyde Ross, and Clyde Meader.

In 1983, four firemen went to the Fire Academy Fire Fighter I & II training in Presque Isle. They were: Nelson Collins, Richard Knight, Larry French, Jr. and Clyde Meader.

In 1983, I served on the state fire training and education committee that was responsible for setting the direction for firefighting training in Maine.

In 1983, dispatching was transferred to the Franklin County Sheriff's Department at the jail.

In 1983, the Squad Truck was replaced with a used ambulance.

The H. K. Porter Extrication Tool was added to reserve equipment in 1983 complete with cutter, spreader, and 3 air bags of differing sizes to enhance our fire-fighting and rescue capability. This tool operates from air bottles and has a ten-ton rating. This is an industrial grade bolt cutter that can turn 50 lbs. of hand pressure into 4,000 lbs. of cutting pressure. This new piece of equipment meant more training on its proper use which was conducted by a factory representative for the Rescue Team. The state reimbursed the town 50% of the cost through a Department of Transportation grant. In the month of November 1983, it was used two times to free people trapped in accidents, and six times in 1984, and frequently thereafter in subsequent years.

In November 1983, I was elected president of the Maine Fire Chief's Association for the year 1984 after having served on the state training and education committee and as vice president of the Maine Fire Chief's Association as well as past president of the Franklin County Fireman's Association. This was the first time a volunteer fire department chief had been elected president of the state association, an honor for me and all in the department. I would later serve on the board of directors of this organization for thirteen (13) years.

In 1984, Firemen S. Clyde Ross and others completed Fire Fighter II training with Basic First Aid and were certified as Fire Fighter II's. Kenneth Desmond, Captain of the Winslow Fire Department, was the Maine Fire Instructor. (Captain Desmond was the 2017 president of the Maine Federation of Fire Fighters.)

In 1985, the First Regional Fire Fighters I and II Schools were hosted by the Farmington Fire Department. One hundred and twenty-five (125) fire fighters attended this event from Franklin County.

On September 3, 1985, Sheila Landry was the first female fire fighter to join the ranks in Farmington. She was EMS certified and a Fire Fighter I candidate. She remained on the force until September 6, 1988 when she retired.

On September 27, 28, 29, 1985 the Farmington Fire Department hosted the 22nd Annual Maine State Federation of Fire Fighters Convention. This was the FIRST for our department. Over 100 fire departments attended from throughout Maine. On Saturday, the 28th, several thousand lined the 3 miles parade route to see 120 units in procession for a parade that lasted 2 ½ hours. The convention was a great success considering the threat of Hurricane Gloria.

During the flood of 1986 the fire department rescued four people by boat stranded at the Farmington Diner and 7-Eleven.

In 1986, extensive in-service training was carried out by our several qualified instructors in Self-Contained Breathing Apparatus (S.C.B.A.), Draft and Relay Pumping, C.P.R. Refresher with practice using the extrication tool and Car Fire Suppression, Ground Ladders and Aerial Ladder uses, Job Safety, Class "B" Burn, Rural Hitch and Fire Ground Survival. The latter three sessions were held in conjunction with our continuing mutual aid training. Forcible entry and fire suppression tactics received much attention this year.

Courtesy of Barbara Yeaton Estate, 1986 Fire Training.

In 1986 a new Poseidon Air Compressor was purchased to fill the Cascade Air System used in filling air bottles for S.C.B.A. (self-contained breathing apparatus). This new compressor replaces an older, outdated model and fills air bottles to the desired pressure quicker and more accurately.

On December 17, 18, 19th, 1986 I served the State Fire Marshal's Office as a member of the Oral Board selecting the best candidates for fire investigations.

On April 1, 1987, the 100 Year Flood caused damage and destruction to Farmington and Franklin County. The fire department was busy with the rescue of people by boat from businesses and homes on Intervale Road. An attempted rescue of a person at Farmington Falls resulted in a near tragedy for two of our Farmington Falls firemen. Numerous cellars had to be pumped out by the fire department, 36 in all.

August 1, 1988 fifty-two (52) firefighters from 10 departments participated in the annual Franklin County Firemen's Association fire attack school which was hosted by our department.

Franklin firefighters train in 100 degree temperatures

By Brad Crafts

FARMINGTON — Temperatures soaring to nearly 100 degrees Saturday created hot conditions for over 50 Franklin County firefighters participating in the annual Franklin County Firemen's Association fire attack school on Saturday. The event was hosted by the Farmington Fire Department.

Farmington Chief Robert McCleery, chairman of the school's committee, said there were 52 firefighters from 10 departments in the training programs. Asst. Chief Clyde Ross, registration and training chairman, welcomed the firefighters and explained the various training classes and work areas. Ross stressed that safety was the prime factor, not only in firefighting but in fire training.

Six classes were held in the near 100 degree temperatures. The house burning in West Farmington sent temperatures soaring up over 100 degrees. The class A control fire burn of the old house and simulated smoke conditions was handled by instructor Clyde Meader, Farmington, assisted by Terry Bell, George Barker, Tim Hardy and Jack Bell.

How to assume and operate fire ground command for fire officers and simulated fires and firefighting procedures was conducted by Tom Keene, Skowhegan.

Forest firefighting techniques and procedures were given to a group of veteran firefighters and hot shot crews by State Forestry Rangers Rudy Davis, Lewis Prescott, Kendall Knowles and Charlie Clukey.

The basic principles of firefighting, handling of hose, ladders, self-contained breathing equipment and various other fire equipment was explained by Richard Knight. He stressed that the first and most important part of a firefighter's work at all times is safety, both to himself and fellow firefighters.

One of the smaller classes, but one

Franklin County firefighters held their annual fire attack school Saturday at Farmington and one of the classes was to control burn an old house in West Farmington. Above the old house roof is in flames and firemen pour water on the fire.

BRAD CRAFTS photo

of the most challenging subjects, was Francis Roderick's pump maintenance and testing group. Roderick, a retired fire inspector and instructor of the State Fire Marshal's office, kept his students analyzing the working of a fire engine pump and hydraulics.

The principles of the pump on a fire truck and the basic operating procedures were explained by Nelson Collins, who completed his day's training

with hands-on pumping from a small pond in a large gravel pit.

Chief McCleery extended his thanks to the instructors, the Forestry Rangers, Asst. Chief Ross, members of the county committee and to the many firefighters who endured the high temperature training and working conditions.

Lunch was served by members of the Farmington Firemen's Auxiliary at the fire station.

Courtesy Franklin Journal, 1988 Attack School

On November 15, 1988, the new "Emergency One" 110 foot aerial ladder truck arrived, better known as "Ladder 1" at a cost of $314,000.00. Upon receipt, training was started immediately and by year end had been used in several situations in surrounding towns.

December 4, 1988 1:00 AM Car Crashes thru Fire Station overhead doors striking Squad Truck head-on pushing it into the aerial trucks. $19,000 in damages. Photo Courtesy Farmington Fire Department.

On December 4, 1988 at approximately 1:00 AM an accident occurred at the fire station, involving a privately-owned vehicle and two (2) fire department vehicles. An individual evading apprehension by the police drove his vehicle through one of the fire station's overhead doors striking the Squad Truck head-on, pushing it back into the old aerial truck, then coming to rest on the front tire of the new ladder truck. This accident caused more than $19,000.00 of damage to fire department property.

In 1989, fire department salaries were $10,525.

In August 1989, a new Ford F 350 Squad Truck was purchased for $30,500.00 to replace the one lost due to the accident at the station in December 1988. The purchase was made possible upon receipt of the insurance money from the accident, appropriations by the town, and by a donation from the Firemen's Benevolent Association. The new Squad Truck included a Darley Pump for $4,000.00, new Honda generator for $1,500.00, and a front winch. In the late 1980's I served as a member of the Franklin County Radio Communications Committee which created county-wide dispatch for all emergency service calls through the sheriff's office, serving 9 towns.

In the 1990's, the "Captain No Burn Program" in the local schools became the "Learn Not to Burn" program which is part of the Farmington (SAD #9) school curriculum for grades K-8 and continues to this day.

On January 9, 1990, Fireman Robert Henry "Apple" Oliver died. He was a member of the Farmington Fire Department for 54 years (1936-1990). You could find "Apple" behind the wheel of a fire truck at most fires.

In 1990, the old aerial ladder truck was converted to a tank truck by our firemen. This truck was a self-contained unit carrying 1800 gallons of water, 1 portable tank, a 300 GPM pump, ladder, generator, and portable pump. Known as "Tank #1". It is a very useful piece of equipment for the Town of Farmington.

In August 1990, the Farmington Fire Department was involved in its first hazardous material spill involving a tank truck loaded with 6000 plus gallons of Black Liquor, an industrial chemical waste of pulpwood processing, whose gases are toxic. Our neighboring mutual aid fire departments responded to assist us, along with International Paper Co., Department of Environmental Protection (DEP), and the transport company's RST Hazardous Material Team. All things considered, the evacuation, salvage and overhaul went well.

Nineteen-ninety (1990) was the beginning of the Haz Mat Program which started with cooperation from several towns. The Fire Department worked with International Paper Co. in Jay and the local departments in the Mutual Aid Network.

In December 1990, the 9-1-1 emergency number went into effect replacing the old 778-2120 fire number. This covered all the 778 prefix numbers and was for all ambulance, fire and police emergencies.

In 1991, the Hazardous Material Team was organized, and in May 1991, the first members of the Farmington Fire Department to be certified as a Hazardous Materials Technician were: Clyde Meader, Tim Hardy, Mike Bell, and George Barker.

23 training sessions were conducted in 1991.

In 1991, the In-House Cascade Breathing Air System was upgraded to 6 bottles, 4500 PSI (pounds per square inch), and the Squad Truck was outfitted with a portable Cascade Breathing Air System of 4 bottles, 4500 PSI. The Cascade System filled 655 air bottles for the Farmington Fire Department and 870 for Mutual Aid Fire Departments.

In 1992, Fire Insurance Classification (ISO-Insurances Services Organization) was completed – Grade 5. The fire department's First Annual Easter Egg Hunt was started in 1992.

Robert "Apple" Oliver - 54 years with FFD.
Where he loved to be - behind the wheel of a fire truck.
Engine 1, 1931 Maxim Pumper. Pictured with Bob McCleery 1975

In 1992 as a past president of the Maine Fire Chief's Association, I served as Director of the Fire Education and Training Committee of that organization. I was also a member of the International Fire Chief's Association, New England Division.

Two (2) Class "A" Burns with Mutual Aid Companies were conducted in 1992. One of the properties for the "Live Fire Training" was located on Lincoln Street.

September 9, 1992, John Edgerly, Farmington Town Manager, nominated me for consideration of the Maine Fire Chief's Association first "Fire Chief of the Year" award. Cape Elizabeth Fire Chief, Phil McGouldrick, was most deservedly selected as the first Maine Fire Chief of the Year. He became a full-time firefighter in 1963, served as Fire Chief in South Portland 1972 - 1992, then Cape Elizabeth Fire Chief 1992 - 2008. He initiated the "Learn Not To Burn" program which would be launched nationwide.

In January of 1993, Terry Bell became the town's Emergency Management Director succeeding Dr. Thomas Eastler, a UMF professor.

In 1993, a new building was constructed at the fair grounds for the Farmington Fire Department Benevolent Association.

In 1993, I attended my first International Association of Fire Chiefs meeting in Dallas, Texas.

1992 Lincoln Street Class "A" Burn Training Chief Robert McCleery

Norman Collins, Assistant Chief 1976 -1993
Photo Courtesy of 1994 Town Annual Report.

In 1994, Assistant Chief Norman E. Collins died. He was a World War II veteran
and a fireman for 46 years. He was a charter member of the Farmington Falls Fire
Department that organized in 1947. When the Farmington Falls Fire Department
became a sub-station of the Farmington Fire Department in 1976, he was named
Assistant Chief.

Twenty-five (25) training sessions were conducted by the department in 1994.

On September 9, 10, 11, 1994, the Farmington Fire Department hosted, for the
second time, the 31st Annual Convention of the Maine State Federation of Fire
Fighters. This three-day event drew one of the largest gatherings of firemen and
fire equipment ever assembled in the state. The parade was one of the longest
with over 280 units, lasting 2 hours, and other events included a bonfire, dance,
and banquet. The Wilton Fire Department coordinated a muster that was a great
success.

In 1994, cellular phones were added to the array of equipment used by firemen.
A truck committee made up of several members of the fire department was busy
writing specs for a new pumper truck.

In 1995, a regional search and rescue unit was established that trained regularly
in cold water, swift water, ice, high and low angle rescues, and also snowmobile
and orienteering rescues.

In July 1995, S. Clyde Ross, Tim Hardy and Mike Bell traveled to Ocala, FL to inspect the new "Emergency One" 1500 gallon pumper truck with 1000 gallon water tank which arrived the first of October. Cost of Truck: $258,000.00.

In 1995, the air compressor was repaired after many years of dedicated service, more 4 inch hose was ordered along with other replacement items, and a new dump tank and rack was put on Engine 2.

In 1996, several firemen attended the Maine State Federation of Firefighters Convention in Ellsworth. Deputy Chief Ross and I attended meetings of the Maine Fire Chief's Association, the New England Association of Fire Chiefs in Marlboro, MA, and the International Association of Fire Rescue in Kansas City.

Search and Rescue & Emergency Training received more attention in 1996. This area of training will continue to be of high significance as more people use our wilderness areas for recreation.

In 1997, I advocated the Department of Transportation for the Hurst "Jaws of Life" hydraulic rescue tool for the Town of Farmington which the department received.

In 1997 discussions took place about the need to replace the Rescue Truck with a new heavy-duty rescue truck to carry all the rescue equipment that was divided between two vehicles. Thanks to a local insurance agent, the fire department had use of a 9-1-1 telephone simulator for use in the elementary grades in the Spring of 1997. The children's response and eagerness to learn was outstanding.

In 1997, several firemen attended the Maine State Federation of Firefighters Convention in Kennebunk, Maine. Deputy Chief S. Clyde Ross and I attended the Maine Fire Chief's Association, the New England Division of the International of Fire-Rescue and the New England Association of Fire Chiefs combined in Springfield, MA. Firemen Terry Bell and Tim Hardy also attended the New England Division of the International Fire-Rescue Convention in Springfield, MA.

January 1998 started off with the ice storm. Power was out for days and weeks in some places; trees and power lines were down throughout the town. Firemen assisted with the removal of trees from roadways and power lines, and helped assist needy citizens to the Community Center Shelter. Some firemen stayed at the shelter and helped with food preparation. Firemen also hauled water with an engine to farms to water livestock.

On February 6, 1998, Harold "Stub" Hemingway was recognized for 56 years of service to the department. He joined the department February 3, 1942.

June 1998 was a month of heavy rain and flooding. Extensive damage was done to roads, fields, and crop land. The firemen patrolled flooded roads and streets in and around Farmington.

July 1998 brought lightning strikes and the MTE, Inc. fire in Fairbanks which was one of the worst fires in Farmington in several years requiring support from several mutual aid fire departments.

In the Summer of 1998, the Farmington Fire Department's Search and Rescue Team was called twice by the Warden's Service to assist with the rescue of injured climbers on Tumbledown Mountain in Weld.

In August 1998 two young firemen, Timothy D. "TD" Hardy and Jason Decker completed Fire Fighter II course held in Waterville. The nine days and evenings of study and field training are conducted by the Maine Fire Service Training and Education.

Deputy Chief Ross and I attended meetings in 1998 of the Maine Fire Chief's Association, and the International Association of Fire Chiefs, New England Division, in Springfield, MA.

At the October 14, 1998 meeting of the Maine Municipal Association's Annual Convention I was once again honored to be one of eleven candidates nominated for 7th Annual Maine State Fire Chief of the Year, and even more humbled to be chosen "Chief of the Year." This award is dedicated to the outstanding Farmington firefighters with whom I have worked and whose hands have helped shape the Farmington Fire Department into the excellent department it is today. (See nomination papers in Appendix.)

In 1998, the fire department received a Life Safety Achievement Award from the Residential Fire Safety Institute presented by Maine State Fire Marshal John Dean for Farmington firefighters' dedication to public safety with no recorded fire deaths in 1998 and the department's long-standing commitment to fire prevention and public education. (This was the 4th year in a row the department had received this award and would again in 1999.)

In 1999, the fire department's training program was continually on-going. These sessions and programs were necessary to keep abreast of changing laws and mandates facing the Fire Service. Today's fire departments are no longer just fire departments: They are Fire, Rescue, and EMS combined into an emergency services department.

In August 1999, the fire department sent two young men, Jon Alexander and Chet Alexander, to the Fire Fighting Academy in Auburn, and others attended one and two-day classes at the regional training sessions in different counties.

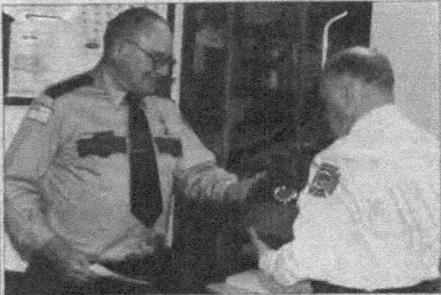

Hemingway CONTINUED FROM PAGE 1

"Stub" Hemingway is recognized for 56-plus years

Recognized - Harold H. "Stub" Hemingway (left) is presented a plaque by Fire Chief Robert McCleery. (Photo by Greg Davis)

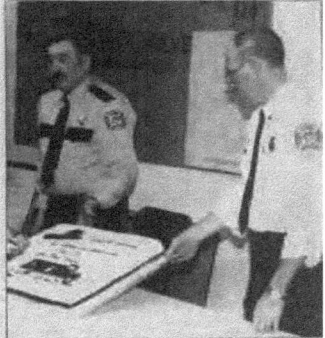

Commemorative Cake - Firemen Eugene Mosher and S. Clyde Ross hold the cake dedicated to their comrade, Harold H. "Stub" Hemingway. (Photo by Greg Davis)

see Hemingway page 12

Harold "Stub " Hemingway Honored - 56 years of service to FFD
February 6, 1998
Courtesy of the Franklin Journal.

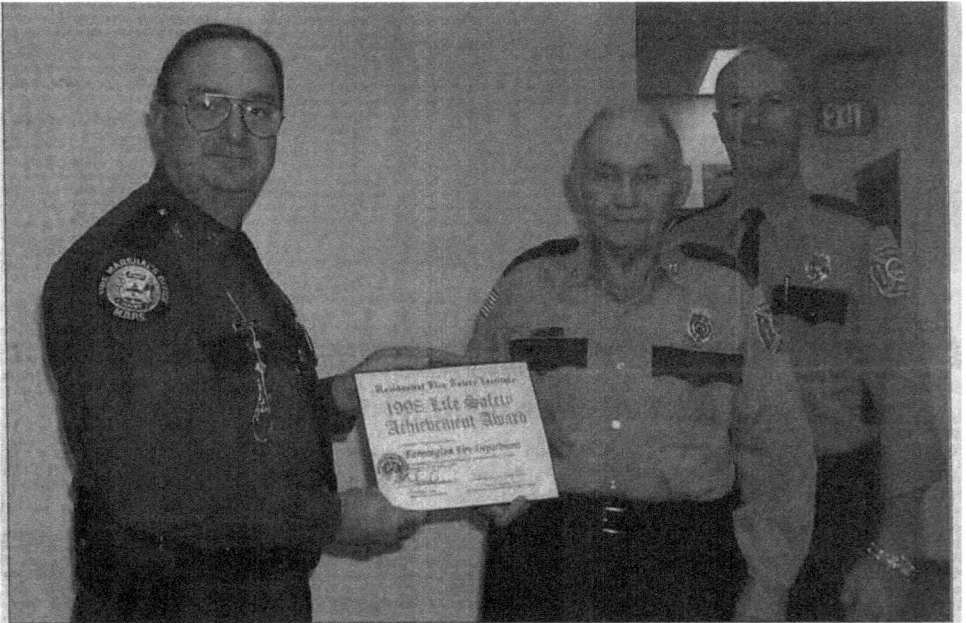

SAFETY ACHIEVEMENT: Maine State Fire Marshal John C. Dean, left, presents Farmington Fire Chief Robert McCleery, center, with a Life Safety Achievement Award from the Residential Fire Safety Institute Tuesday night. Farmington's Deputy Fire Chief S. Clyde Ross is at right. The award is given to the department because it recorded no fire deaths in 1998. While 20 percent of all U.S. fires are residential in nature, they contribute to 80 percent of all fire related deaths. McCleery turned the award over to Town Manager Pamela Corrigan, who said it would be put on display in the Town Office. Dean said the award showed the dedication Farmington firefighters have when it comes not only to responding to fires but also in fire prevention and public education.

1998 Life Safety Achievement Award presented to the Fire Department by
State Fire Marshal John Dean on left.

In 1999, an exhaust system was installed in the apparatus bay to exhaust the toxic fumes from the trucks when started. The system makes the air cleaner providing a healthier work place for the whole building. Nineteen-ninety-nine (1999) completes over a decade of being dedicated to, and focused on, training within the Farmington Fire Department. It has been the practice of this department, and its leadership, to upgrade the skill levels of firemen and introduce new techniques and methods for meeting emergency situations. We continue to meet the legislature's mandated standards in the statutes and new pieces of legislation as it relates to emergency response. In 1999, the fire department was managed by a part-time, on-call chief with officers who received annual stipends and hourly wages according to their responsibilities. Stipends and wages were also paid to firefighters based on training, experience, and availability to respond to emergency calls.

On January 18, 2000, I tendered my resignation as Fire Chief effective July 1, 2000 after 23 years as chief and 40 years as a member of the department. I left the department in very capable hands.

* On June 26, 2000 Terry S. Bell, Sr. became Farmington's first full-time Fire Chief with Deputy Chiefs S. Clyde Ross and Timothy A. Hardy; Senior Captains: John O. Bell, Morrill W. Collins and Harold H. Hemingway; Captains: Michael A. Bell, Nelson E. Collins and Richard A. Knight; Lieutenants: George L. Barker and Clyde H. Meader. Firefighters: Chet C. Alexander, Jonathan M. Alexander, Philip R. Allen, Melvin L. Bard, Peter H. Brennick, James A. Brown, Stephen M. Bunker, Richard A. Chabot, Jonathan E. Davis, Jason A. Decker, David M. Fronk, Timothy D. Hardy, Rocky Jackson, Curtis C. Lawrence, Robert E. McCully, Eugene L. Mosher, Douglas R. Oliver, Raymond G. Pillsbury, David A. Pottle, Donald A. Richard, Gregory E. Roux, Junior E. Turner and Randall A. Voter; Rookies: Stephen LeTarte, Dean Frost and Jason Hyde; and Chaplain Stan Wheeler. Within above group are 19 Fire Fighter II, 6 Fire Fighter I, 5 Certified Instructor I who guide other firemen in the fire department, a Hazardous Material Response Team, 3 Emergency Medical Technicians, 1 Paramedic and 1 Emergency Medical Technician Intermediate.

1999 FFD Officers: Assistant Chief Tim Hardy, Deputy Chief S. Clyde Ross, Chief McCleery, Deputy Chief Terry S. Bell, Sr.

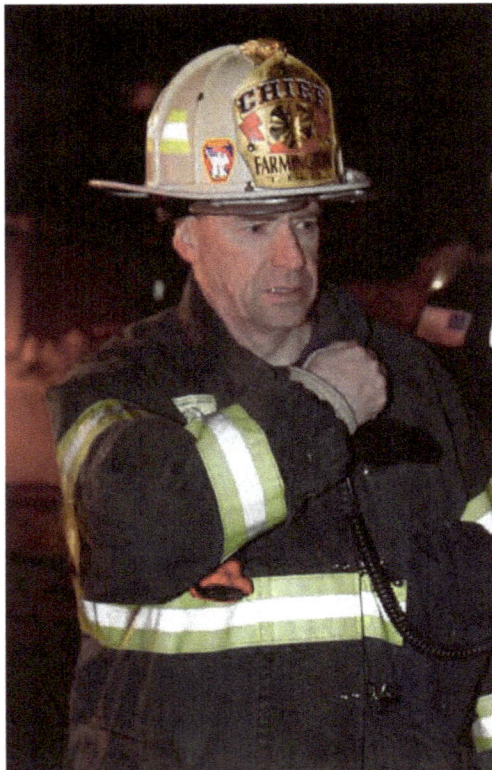

Chief Terry S. Bell, Sr. 2000 - Present

Chapter 4

FARMINGTON FIRE DEPARTMENT VEHICLE PURCHASES

I will list as near as possible the fire engines purchased 1860 - 1955:

1860	Hunneman Rebuilt 1836 Hand Pump fire engine
1887	Horse drawn Steam Engine
1912	Ford Pumper
1916	Model T Chemical fire truck
1925	Dodge Graham hose and ladder truck
1931	Maxim Pumper 500 GPM, "Engine 1"
1948	Chevrolet 500 gallon water tanker
1955	Ford F 750 fire truck 500 GPM front mount pumper with 750 gallon water tank

Below is a list of vehicles added to the fleet from 1966 through 1995:

1966	Ford Thibault 750 GPM pumper from Tennessee with 750 gallon water tank, "Engine 4"
1975	Mack Pumper 1000 GPM with 750 gallon water tank. To be known going forward as "Engine 5"
1977	The first Squad Truck was purchased
1979	Ford C 900 truck with used 75 foot ladder was added
1982	Added a second Mack truck, 1000 GPM with 750 gallon water tank, "Engine 6"
1983	Replaced the Squad Truck with used ambulance from Hopkinton, MA
1988	New 110 foot "Emergency One" ladder truck, "Ladder 1"
1989	Purchased a new Ford F 350 Squad Truck with winch, Darley Portable Water Pump & Honda Generator for fire scene lighting
1990	Converted old ladder truck to an 1800 gallon tank truck, "Tank 1"
1995	Purchased an "Emergency One" 1500 GPM pumper truck with 1000 gallon water tank

It is with gratitude to Fire Chief Terry Bell and Deputy Chief S. Clyde Ross that the following equipment purchases can be added to this list since Chief McCleery's retirement and passing:

2000	Chevrolet 4 X 4 Pick-Up Truck for First Full-Time Chief's Use
2003	Pierce Rescue Pumper, 1500 GPM pump with 750 gallons of water, "Engine #1"
2004	Chevrolet Suburban "COMMAND" Communication Vehicle, "SUV"
2007	Pierce Tower Ladder 100 foot with 2000 GPM pump & 400 gallons of water, "Tower 3"
2010	Ford F 150 4 x 4 Chief's Truck
2012	Ford F 550 4 x 4 Squad Truck, "Squad #1"
2015	John Deere Gator with Wheels and Track System

OVERVIEW OF FIRE ENGINE COSTS 1860 - 2007

1860	1886	1954	1965	1974	1981	1983	1995	2003	2007
$600	5,000	9,000	15,000	53,250	106,000	314,000	258,500	369,500	812,000

SQUAD TRUCK COSTS 1977 - 2012

1977	1989	2012
$ 25,000	30,500	94,000

1860 Hunneman "Eagle 1" Hand Pump Fire Engine Photo Courtesy of Henry Ford Museum

1916 Model T Chemical Fire Engine Photo Courtesy of Owls Head Transportation Museum

Chief Vic Huart

1931 Maxim Pumper, Engine 1, Duane Hardy standing with arm on windshield, others unidentified, in Fairbanks.

1931 Maxim Pumper 500 GPM, Engine 1

Mack Pumper 1000 GPM, Engine 5, Delivered 1975 Pictured: Chief Forest Allen, Cy Decker, Bob McCleery

1977 Fire Engines L-R: Engine 4, Richard Russell; Engine 5, Cy Decker; 1955 Ford F 750 Front Mount Pumper, Clyde Ross; 193' Maxim Pumper, Engine 1, Guy Brann

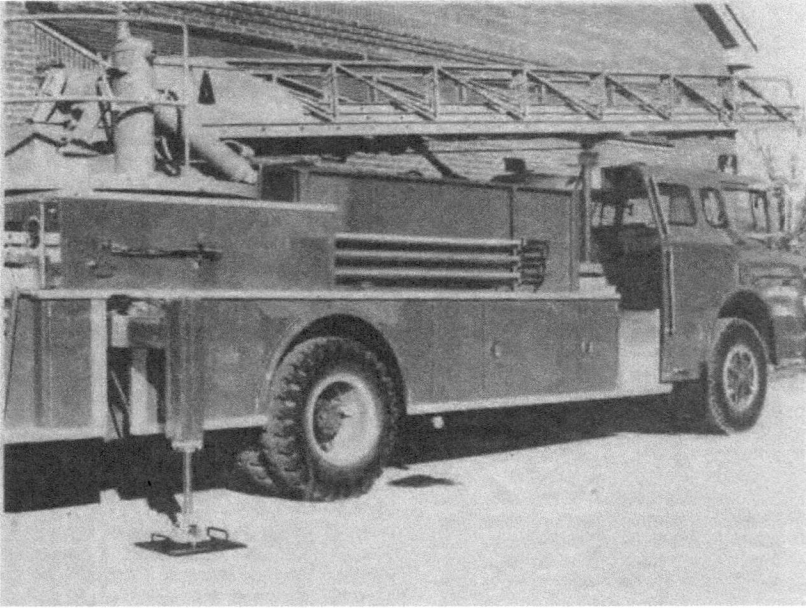

1979 Ford Aerial Truck Photo Courtesy Barbara Yeaton

1982 Mack Truck, Engine 6, L-R: Richard Russell, Chief McCleery, Doug Oliver, Tim Hardy

Pictured L-R: 1985 Fire Engines: Squad 1, Engine 1, Engine 3, Engine 4, Engine 5, Engine 6, Original Ladder 1

1988 110 ft. Emergency One, Ladder 1, Chief McCleery

1989 LADDER 1 COMPANY 1st Row L-R: James Brown, TD Hardy, Richard Chabot 2nd Row L-R: Jon Davis, Melvin Bard, Chet Alexander, Richard Knight, Jon Alexander, Jason Decker, David Pottle

1995 Emergency One, Ladder 1

1999 FIRE ENGINES L-R: Ladder 1, Engine 2, Engine 6, Tanker 1, Squad 1.
(1990 FFD firemen converted older ladder truck to 1800 gal. tank truck)

All FFD Apparatus & Members Photos Courtesy of Farmington Fire Department and Town of Farmington

2014 Fire Engines

Squad 1
2012 Ford F550 squad/brush unit, with 250 GPM pump,
200 gal skid tank with 5 gallons foam

Engine 2
1995 E-One Cyclone Pumper.
1500 GPM pump, 1000 gal tank

Engine 1
2003 Pierce Dash Resuce Pumper 1500 GPM pump,
750 gal tank

Tower 3
2007 Pierce Dash 100' Tower Ladder, 2000 GPM pump,
300 gal tank, and 30 gal foam tank

C-1
2010 Ford F150 Crew Cab, Chief's Truck

Chapter 5

FIRE DEPARTMENT MEMBERS

I do not have a list of Farmington Fire Department members prior to 1885, with the exception of Captain Jennings who led the fire department in 1860, and Captain F. V. Stewart and Captain E. I. Merrill in 1873. I am assuming the name at the head of the list was either the chief, foreman or captain of the department. In listing them, I will assume the first name on the list served in the capacity of Fire Chief. There were no distinctions made in listing the assistant chiefs on the membership rolls until 1951 when the chief's officers were named in order of rank: chief, 1st assistant chief, 2nd assistant chief, followed by secretary-treasurer.

Prior to 1923 the chief officer changed frequently. In 1923, Victor C. Huart was elected Fire Chief where he served until his death at the age of 66 on March 7, 1951. Chief Huart served for 28 years, and was the Farmington Fire Department's longest serving chief. The shortest tenure of any chief, was J. Ambrose Compton who in 1951 was elected FireChief and immediately resigned. He was chief for 15 minutes. James Bauer Small was then elected Fire Chief and served from April 2, 1951 until retiring December 2, 1969, 18 years as chief. Forest L. Allen was elected Fire Chief December 2, 1969 and served until retiring March 31 ,1977, 8 years as chief. On April 1, 1977, I, Robert L. McCleery, was appointed Fire Chief and served until retiring effective July 1, 2000, 23 years as chief. Terry S. Bell, Sr. became Farmington's first full-time Fire Chief. I had advocated for years for the town's need to hire some full-time firemen as had my predecessors Chief Bauer Small and Chief Forest Allen. I was pleased that in the new millennium a new chapter had begun for the Farmington Fire Department with the appointment of its first full-time Fire Chief. I will be awaiting the appointment of other full-time firemen in the years to come.

Members of the Fire Department

1860   Captain Jennings

1873   Captain F. V. Stewart
        Captain E. I. Merrill

1886   E. I. Merrill, Foreman (Chief) – until 11/26
        George C. Purington, Foreman (Chief) 11/26
        I.Warren Merrill, First Assistant Foreman

Other members of the Farmington Steam Fire Company No. 1 were:

W. B. Elwell, 2nd Asst. Foreman; E. E. Jennings, 3rd Asst. Foreman; Chas. E. Wheeler, 1st Engineer; Chas. W. Marston, 2nd Engineer; Ruben Huse, First Fireman; C. A. Priest, 2nd Fireman; Arthur M. Bailey, First Pipe Man; Frank A. Davis, First Pipe Man; L.M. Burbank, Suction Hose Man; F. P. Adams, Suction Hose Man; and Firemen W. B. Bailey, Samuel Bailey, E. W. Bragg, Hannibal Berry, J. A. Blake, Fred A. Collins, F. E. Doughton, J.M. Matheau, Carl Merrill, Dana Merrill, Stephen W. Carr, Joseph Pero, Geo. McL. Presson, Joseph Pero, John Robinson, and George E. Stevens.

1887	E. E. Rufus Jennings
	J. A. Blake

1888	E. E. Rufus Jennings

1890	Carl P. Merrill
	Joseph Matheau

1891	W. B. Elwell
	J. M. Matheau
	A. W. True

1892	W. B. Elwell
	A.W. True
	Edwin A. Higgins

1898	Carlton P. Merrill
	E. A. Odell
	H. J. Spinney

1900	E. M. Higgins
	J. M. Soule
	J. M. Matthew

1901	C. B. Moody
	J. M. Matheau
	Rufus Jennings

Farmington Steam Fire Engine Company No.1.

1902 J. M. Matheau   Chief
      C. B. Moody
      J. M. Soule
      C. H. Brimmer
      E. M. Higgins
      Henry Knapp
      H. C. Spinney
      E. E. Rufus Jennings
      F. W. Lawry
      Elmer Weymouth
      C. P. Morrill
      Leo Russell
      J. A. Blake
      Benjamin Hayes
      Geo. Blake
      J. H. Gilkey
      R. Campbell
      J. A. Linscott

1903	J. A. Blake	Dropped
	Robert Campbell	Joined
	Willis Cook	Joined
	Leo Russell	Dropped 3/5
	Henry Knapp	Resigned 11/3
1904	F. W. Lawry	Dropped 2/2
	Roderick Archer	Joined
	J. M. Matheau	Chief Resigned 3/1
	C. B. Moody	Dropped
	Geo. Blake	Chief 5/1
	Joe Chick	Joined
	Forest Lock, Jr.	Joined
	E. E. Flood	Joined
	Earl Milliken	Joined
	C. H. Brimmer	Dropped
	Harry Wheeler	Resigned 12/6
	H. C. Spinney	Resigned
1905	C. B. Moody	Joined 1/1
	Arthur Keith	Joined
	Robert Campbell, Jr.	Resigned 10/13
	Frank Fuller	Joined
1906	Arthur Tucker	Joined
	J. M. Soule	Resigned 11/6
1907	Leon H. Marr	Joined
1908	Lewis Marsh	Joined
	E. E. Flood	Chief
	H. Spinney	Joined
1909	Arthur Tucker	Chief
	William Flood	Joined 1/1
1910	E. B. Kempton	Chief
	C. B. Moody	Resigned 5/10
	H. E. Spinney	Resigned
	Bert S. Pratt	Resigned
	Geo. McLeod	Joined 5/10
	E. D. Jackson	Joined
1911	E. B. Kempton	Chief
	Arthur Tucker	Dropped
	A. B. Carr	Joined 4/1
1912	A. B. Carr	Chief
	E. E. Flood	Resigned 3/5
	Forest Locke	Dropped 4/19
	A. E. Roderick	Dropped
	E. W. Milliken	Resigned 3/5
	Geo. McLeod	Dropped

	David Boone	Dropped	
	John Buzzell	Joined	
	C. B. Moody	Joined	
	Ralph Morton	Joined	
	Thomas Gagne	Joined	
1913	Charles Nickerson	Joined 1/1	
	Fred E. Stinchfield	Joined	
1915	Forest Locke	Joined	
	Bert L. Voter	Joined	
	Albert Jackson	Dropped	
	A. B. Carr	Chief	Resigned
1916	A. D. Keith	Chief	
	Harry Koch	Joined	
	H. H. Hinds	Joined	
	Carroll L. Whitney	Joined	
	Joe Chick	Dropped	
1917	E. D. Jackson	Chief	
	Geo. E. Peachey	Joined 12/5	
1918	Victor C. Huart	Joined 3/5	Would become Chief 1923
1919	John Buzzell	Resigned 4/1	
	Forest Locke	Resigned 5/6	
	Raymond T. Currier	Joined	
	Archer Kidder	Joined 6/3	Resigned 11/4
1920	Frank Craig	Resigned 1/6	
	Geo. Harry Koch	Chief	
	Bertram Starbird	Joined 5/4	
	Roy Stinchfield	Joined 5/4	
1922	Cloud Jordan	Joined 7/1	
1923	Victor C. Huart	Elected Chief 1/1	
	Elden Hall	Joined 1/2	
	Carl H. Fenderson	Joined 7/1	
	E. C. Gray	Joined 12/4	
1925	Geo. W. Dobbins	Dropped 2/3	
	Raymond T. Currier	Resigned 3/3	
	Roy E. Hobbs	Joined 3/3	
	Henry Greaton	Joined 3/3	
	Forest P. Whittier	Joined 7/7	
1926	Walter D. Barker	Joined 4/6	
	Robert Campbell	Died 3/10	
	Elmer Weymouth	Resigned 12/7	
	Fred Stinchfield	Resigned 12/7	
	Dan T. Adams	Joined 12/7	

1927	Ed. H. Connors	Joined 3/1	
	Arthur Compton	Joined 9/6	
1928	Carroll L. Collins	Joined 2/7	

Farmington Fire Company No. 1

1929	Victor C. Huart	Chief	
	J. H. Gilkey	Retired 1/1	
	Wilbur Smith	Joined 5/5	
	J. Ambrose Compton	Joined 10/1	
1931	Ralph Morton	Retired 3/3	
	Gardner Parlin	Joined 3/3	Rookie
	Rockwell Flint	Joined 3/3	Rookie
1934	Leon Tardy	Joined 12/4	Rookie
	Dan Adams	Retired 12/4	
1935	Addison Linscott	Joined 1/1	Rookie
	Wilbur Smith	Retired 1/1	
	Elden Hall	Retired 12/3	
	Richard Oliver	Joined 12/3	Rookie
1936	Frank M. Deering	Joined 1/7	Rookie
	Robert Oliver	Joined 3/3	Rookie "Apple" Oliver
	J. Bauer Small	Joined 4/7	Would become Chief 1951
	Arthur Compton	Retired 8/4	
	Maurice Taylor	Joined 12/1	Rookie
1938	Bert L. Voter	Retired 12/6	
	Phil Folger	Joined 1/3	Rookie
1939	Richard Oliver	Active 6/6	
1940	P. C. Dennison	Joined 2/6	Rookie
	Addison Linscott	Retired 6/4	
1941	George L. Hobbs	Joined 8/5	Rookie
	Roy L. Weston	Joined 8/5	Rookie
	Carroll Sinskie	Deferred	
	Rockwell Flint	Died – Oct.	
1942	Carroll Collins	Retired 1/6	
	Donald Mosher	Joined 1/6	Rookie
	Norman Sawyer	Joined 1/6	Rookie
	A. Thomas Clark	Joined 1/6	Rookie
	H. H. Hemingway	Joined 2/3	Rookie "Stub" Hemingway
	C. G. Nickerson	1st Asst. Chief 5/5	6 Mo Leave of Absence
	Roy Stinchfield	1st Asst. Chief 5/5	
	J. Bauer Small	Clerk 5/5	
	George Kershiner	Joined 6/2	Rookie
	Atherton F. Ross	Joined 6/2	Rookie

	Clyde L. Richardson	Joined 6/2	Rookie	
	Roy Hobbs	Dropped 7/7		
	Arthur Cutler	Joined 11/3	Rookie	
1943	Clyde Richardson	Resigned 1/5		
	Percy Dennison	Dropped 1/5		
	Roy Weston	Dropped 1/5		
	Henry Greaton	Dropped 1/5		
	Phil Folger	6 Mo Leave of Absence 4/6		
	Charlie Barker	Joined 4/6	Rookie	
	Phillip Bacon	Joined 4/6	Rookie	
	Delmar Johnson	Joined 4/6	Rookie	
	Fred Sturtevant	Joined 4/6	Rookie	
	Wesley Bacon	Joined 4/6	Rookie	
	Richard Oliver	6 Mo Leave of Absence 5/4		
	Roy Weston	Joined 7/6	Rookie	
	Sayward Ross	Joined 8/3	Rookie	
	Richard Oliver	Placed on Honorary List 11/2		
	A. Thomas Clark	Called to Military Service 11/2		
	Forest Allen	Joined 12/7	Would become Chief 1969	
1944	Arthur Locke	Retired 2/2		

Officers Elected:

V. C. Huart	Chief		
J. Bauer Small	Sec.-Treas.		
Roy Stinchfield	1st Asst. Chief		
Elvet Gray	2nd Asst. Chief		

Henry Greaton	Joined 4/4		
Atherton Ross	Retired 5/2		
Maurice Taylor	Joined 5/2	Rookie	
Norman Sawyer	Leave of Absence 9/5		
Walter Barker	Retired 12/5		

1945	H. H. Hemingway	Military Service 2/6	"Stub" Hemingway
	Kenneth Trask	Joined 3/6	
	George Hobbs	Leave of Absence 3/6	
	George Hobbs	Returned 4/3	
	Charles Barker	Retired 4/3	
1946	Richard Oliver	Re-Joined 5/7	
	Bertrum Starbird	Died 4/25	Bertrum "Mul" Starbird
	Carroll Sinskie	Joined 10/1	Rookie
	Phil Folger	Joined 10/1	Rookie
	George Kershner	Died – Nov.	
	Roy Stinchfield	Retired 12/3	26 yrs. svc.

1947	Gardner Parlin	Retired 1/7	
	James Waugh	Joined 1/7	Rookie
	Harley Oliver	Joined 2/4	Rookie
	Phillip Bacon	6 Mo Leave of Absence 2/4	
	Norman Sawyer	6 Mo Leave of Absence 5/6	
	Harrison Bragdon	Joined 4/9	1st Training Officer
	Duane Hardy	Joined 6/3	Rookie
	Phillip Bacon	Retired 8/5	Honor Roll 9/2
1948	Elvet Gray	Retired 12/7	24 yrs. svc.
1948	Officers Elected:		
	Victor C. Huart	Chief	
	J. Ambrose Compton	1st Asst. Chief	
	J. Bauer Small	2nd Asst. Chief	
	Stewart Whittier	Joined 2/1	Rookie
1950	Henry Greaton	Retired 1/3	
	Howard White	Joined 8/1	Rookie
1951	Victor C. Huart Chief	Died 3/7 Age 66	Chief 23 yrs.
	J. Ambrose Compton	Elected Chief 4/2,	Resigned as Chief 4/2
	J. Bauer Small	Elected Chief 4/2	

Officers Elected 4/2:

	J. Bauer Small	Chief	
	J. Ambrose Compton	1st Asst. Chief	
	Delmar Johnson	2nd Asst. Chief	
	Duane Hardy	Sec.-Treas.	
	Earl Knapp	Joined 4/3	Rookie
	Richard Childs	Joined 5/1	Rookie
1953	Sayward Ross	Retired 7/7	
	Tom Hart	Joined 8/4	Rookie
1954	J. Ambrose Compton	Retired 4/6	25 yrs. svc.
	Pete Durrell	Joined 4/6	Rookie
	Maurice Taylor	2nd Asst.	Chief 4/6
	James Waugh	Retired 11/2	
	Clifford Neil	Joined 11/2	Rookie
1956	Tom Hart	Resigned 3/6	
	George Yeaton	Joined 4/2	Rookie
	Harley Oliver	Retired 5/1	
	Bernard Thomas	Joined 5/1	Rookie
1957	Tom Hart	Joined 2/5	Rookie
	Leon Hardy	Retired 4/2	
	Woodrow W. Adams	Joined 7/2	Rookie
	Carroll Sinskie	Retired 9/3	

	Duane Hardy	Retired 10/1	
	Kenneth Higgins	Joined 11/5	Rookie
	Arthur Cutler	Sec.-Treas. 11/5	
	Clinton B. Blaisdell	Joined 12/3	Rookie
1958	Carl Durrell	Joined 5/6	Rookie
	Richard Childs	Retired 9/2	
1959	Officers Elected:		
	J. Bauer Small	Chief	
	Maurice Taylor	1st Asst. Chief	
	Forest Allen	2nd Asst. Chief	
	Arthur Cutler	Sec.-Treas.	
	Delmar Johnson	Retired 12/1	
	Richard Oliver	Retired 12/1	
1960	Stewart Whittier	Retired 1/3	
	Robert L. McCleery	Joined 2/2	Would become Chief 1977
	John Bell	Joined 3/1	Rookie

Town takes over Fire Dept. from Farmington Village Corp. March 10, 1960

1961	Richard Oliver	Died	Funeral 2/27 15 yrs. svc.
	Tom Clark	Dropped 12/5	
	Kenneth Trask	Retired 1/3	
	Wesley Bacon	Retired 2/7	
	Richard A. Russell	Joined 2/7	Rookie
	Ivan L. Howard	Joined 2/7	Rookie
	Woodward Adams	Retired 9/5	
	Earl Knapp	Retired 12/5	
	Robert Marquis	Joined 12/5	Rookie
	Maynard Towle	Joined 12/5	Rookie
1962	Harvey Smith	Joined 1/2	Rookie
	Charles Parlin	Joined 8/7	Rookie
1963	Charles Parlin	Resigned 5/7	
	Douglas Oliver	Joined 5/7	
	Tom Hart	Resigned 12/3	
1966	Wayne Walker	Joined 2/1	Rookie
	Kenneth Durrell	Joined 4/5	Rookie
1967	Officers Elected:		
	J. Bauer Small	Chief	
	Forest Allen	1st. Asst. Chief	
	Harrison Bragdon	2nd Asst. Chief	
	David Ferrari	Joined	Rookie
	Maurice Taylor	Retired 12/6	

1968	Arthur Cutler	Retired 12/3	
	Clifford Neil	Sec.-Treas. 12/3	
1969	Clyde Meader	Joined 2/4	Rookie
	James P. Dunn	Joined 2/4	Rookie
	George Yeaton	Retired 4/1	
	Pete Durrell	Resigned 11/4	
	Gerald Cookson	Deferred 11/4	
	J. Bauer Small Chief	Retired 12/2	Chief 18 yrs.
	Forest L. Allen	Elected Chief 12/2	
	Phil Folger	Retired 12/2	31 yrs. svc.
1970	Pete Durrell	Re-Joined 1/6	
	Gerald Cookson	Joined 1/6	Rookie
	Glenwood Farmer	Joined 4/3	Rookie
	Guy Brann	Joined 6/2	Rookie
	Charles Grant	Joined 9/1	Rookie
1971	James Dunn	Retired 5/4	
	Junior Turner	Joined 5/3	Rookie
	S. Clyde Ross	Joined 5/4	Rookie
1972	Duane Hardy	Died, Funeral 4/25	
1974	Donald Hinckley	Dropped 9/3	
	Terry Warren	Joined 9/3	Rookie
	Clifford Neil	Retired 12/3	
	Carl Durrell	Retired 12/3	
	Robert L. McCleery	Sec. – Treas. 12/3	
	Thomas Cassidy, Jr.	Joined 12/3	Rookie

Farmington Fire Dept. became the Town of Farmington Municipal Fire Department Feb. 10, 1975

1975	Roger Ladd	Joined 4/4	Rookie	
1976	Harrison Bragdon	Asst. Chief	Died 12/15	29 yrs. svc.
1977	Forest L. Allen	Chief	Retired 3/1	Chief 8 yrs.
	Glenwood Farmer	Deputy Chief 3/3		Training Officer
	Robert L. McCleery	Elected Chief 4/1		
	Richard Russell	Sec. – Treas. 4/5		
	Scott Adams	Joined 3/1	Rookie	Retired 6/7
	Terry S. Bell, Sr.	Joined 4/5		Became Chief 2000
	Mike Brinkman	Joined 6/7	Rookie	
1978	George Hobbs	Asst. Chief	Retired 1/3	33 yrs. svc.

Officers Elected 2/6:

	Robert L. McCleery	Chief	
	S. Clyde Ross	Asst. Chief	
	H. H. Hemingway	Captain Harold "Stub" Hemingway	
	John (Jack) Bell	Captain	

Clyde Meader	Lieut.	
Timothy "Tim" Hardy	Joined 3/7	Rookie
Cyrus Decker	Retired 4/4	
Melvin Bard	Joined 4/4	Rookie
Paul Karkos	Joined 4/4	Rookie
Maynard Towle	Retired 5/2	
Mike Knudtson	Joined 5/2	Rookie
Greg Wilson	Joined 5/2	Rookie
Clinton Blaisdell	Lieut. 7/11	
Terry Warren	Retired 12/5	
Steve Bunker	Joined 12/5	Rookie

1979

Guy Brann	Retired 2/2	
Philip R. Allen	Joined 3/6	Rookie
Jim Grant	Joined 3/6	Rookie
Leroy Hazzard	Retired 6/12	
Nelson Collins	Joined 6/12	Rookie
Mike Knudtson	Retired 8/8	
Howard White	Retired 9/11	
Donald Richard	Joined 9/11	Rookie
Richard Knight	Joined 10/9	Rookie

1980

Jed Chase	Joined 1/8	Rookie
Brian Lister	Joined 1/8	Rookie
Kenneth Durrell	Lieut. 1/8	
Morrill Collins	Lieut. 1/8	
Mike Brinkman	Retired 8/12	

1981

Gerald Cookson	Retired 1/13
Charles Grant	Retired 11/10
Pete Durrell	Retired 12/8

1982

James Kiernan	Joined 1/12	Rookie
David Ferrari	Retired 7/13	
Brian Lister	Retired 10/12	
Peter Brennick	Joined 10/12	Rookie
Clinton "Kitty" Blaisdell	Retired 12/7	

1983

Randy Voter	Joined 1/11	Rookie	
Larry French, Jr.	Joined 1/11	Rookie	
Terry Bell	Lieut. 1/11		
Mahlon Moore	Lieut. 1/11		
Richard Russell	Asst. Chief	Retired 11/8	22 yrs. svc.

1984

Terry Bell	Asst. Chief & Sec. 1/10	
Steve Bunker	Treas. 1/10	
Tim Hardy	Lieut. 8/14	
Nelson Collins	Lieut. 10/9	
Jed Chase	Retired 6/12	
Mahlon Moore	Retired 8/14	
Mike Kiernan	Joined 8/14	Rookie
George Barker	Joined 8/14	Rookie

	Eugene Mosher	Joined 9/11 Rookie	
	Jim Grant	Retired 11/13	
1985	Raymond Pillsbury	Joined 1/28	Rookie
	James Sawyer	Retired 7/2	
	Steve Bryant	Joined 7/2	Rookie
	Ken Landry	Joined 9/3	Rookie
	Sheila Landry	Joined 9/3	Rookie
1986	Paul Karkos	Retired 1/7	
	Jim Brown	Joined 3/4	Rookie
	Ivan Howard	Retired 4/1	
	Jed Iverson	Joined 4/9	Rookie
	Richard Knight	Resigned 5/6	Reinstated 12/2
1987	Clyde Meader	Treas. of Benevolence Assn.	
	Roger Ladd	Retired 7/7	
	Greg Wilson	Retired 9/1	
	Art Rogers	Retired 10/6	
1988	Richard Knight	Lieut. Ladder #1 6/7	
	Kenny Dunham	Joined 1/5	Rookie
	Mike Melville	Joined 1/5	Rookie
	Dick Chabot	Joined 1/5	Rookie
	Rocky Jackson	Joined 1/5	Rookie
	Larry French	Retired 1/5	
	Tom Cassidy	Retired 1/5	
	Kenneth Durrell	Retired 4/5	
	Kenneth Landry	Retired 9/6	
	Sheila Landry	Retired 9/6	
1989	David Pottle	Joined 1/3	Rookie
	Tom Savage	Joined 1/3	Rookie
	Mike Bell	Joined 3/7	Rookie
	Kenny Dunham	Retired 6/6	
	Mike Melville	Dropped 12/9	
1990	Robert H. Oliver	Died 1/9	49 yrs. svc. "Apple" Oliver
	Larry French, Jr.	Reinstated 4/3	
	Jim Kiernan	Retired 7/9	
	Tom Savage	Retired 9/4	
	Larry French, Jr.	Resigned 9/4	
	Mike Bell	Treas. of Benevolence 12/8	
1991	Steve Hall	Joined 1/2	Rookie
	Jason Moen	Joined 1/28	Rookie
	Robert Tarbox	Joined 1/28	Rookie
	David Fronk	Joined 4/2	Rookie
	Glenwood Farmer	Deputy Chief	Retired 8/6 21 yrs. svc.
	S. Clyde Ross	Promoted to Deputy Chief 8/6	
	Terry Bell	Promoted to Deputy Chief 8/6	

	Tim Hardy	Promoted to Asst. Chief 8/6	
	George Barker	Promoted to Lieut. 8/6	
1992	Harvey Smith	Inactive 11/3	
	Bradford Moore	Inactive 11/3	
	Jon Davis	Joined 12/5	Rookie
1993	Norman Collins	Asst. Chief	Retired 3/3 46 yrs. svc.
	Morrill Collins	Promoted to Capt. 3/3	
	Robert Tarbox	Resigned 6/6	
	Jason Moen	Resigned 6/6	
	James Dock	Joined 11/2	Rookie
1994	Curtis Lawrence	Joined 2/1	Rookie
	Greg Roux	Joined 11/1	Rookie
	Stan Wheeler	Appointed Chaplain 11/1	
1995	Jed Iverson	Retired 1/3	
	Mike Kiernan	Retired 1/3	
	Steve Hall	Retired 1/3	
	David Pottle	Appointed Benevolence Sec.	
1996	Jason Decker	Joined 9/3	Rookie
	James Dock	Resigned 10/24	
1997	Timothy D. Hardy	Joined 7/1	Rookie

1998 **Officers of the Farmington Fire Department**

Robert L. McCleery	Chief
S. Clyde Ross	Deputy Chief
Terry S. Bell	Deputy Chief
Timothy Hardy	Asst. Chief
John O. Bell	Captain
Morrill W. Collins	Captain
Harold H. Hemingway	Captain
George L. Barker	Lieut.
Michael A. Bell	Lieut.
Peter H. Brennick	Lieut.
Nelson E. Collins	Lieut.
Richard A. Knight	Lieut.
Clyde H. Meader	Lieut.
Stanley Wheeler	Chaplain

Firemen:

Jonathan M. Alexander	Joined 8/4	Rookie
Chet C. Alexander	Joined 11/3	Rookie
Philip R. Allen		
Melvin L. Bard		
James A. Brown		
Stephen M. Bunker		
Richard A. Chabot		

Jonathan E. Davis
Jason A. Decker
David M. Fronk
Marc Hand   Joined 11/3   Rookie
Timothy D. Hardy
Rocky Jackson
Curtis C. Lawrence
Robert E. McCully
Eugene L. Mosher
Douglas R. Oliver
Raymond G. Pillsbury
David A. Pottle
Donald A. Richard
Gregory E. Roux
Randall A. Voter
Junior E. Turner

1999   **Officers and Firemen same as 1998**
Marc Hand   Dropped 4/1

1999   **Members of Ladies Auxiliary**
Sharon Barker   President
Linda Brown   Vice President
Cheryl Bunker   Sec. & Treas.
Sandy Knight   President Emeritus
Eletrice Farmer
Rachel Fronk
Jennifer Lawrence
Melissa LeTarte
Jamie Sullivan
Loretta Bard
Melissa Allen
Lois Bubier
Linda Chabot
Sara MaKenon
Glenwood Farmer   Honorary Member
Florence Norton King   Forerunner to Group 1951 - 1969

2000   **Officers of Farmington Fire Rescue**
Robert L. McCleery   Chief   Retired Effective 7/01   Chief 23 yrs.
Terry S. Bell, Sr.   Elected Chief 6/26/2000
S. Clyde Ross   Deputy Chief
Timothy A. Hardy   Promoted to Deputy Chief
John O. Bell   Promoted to Senior Captain
Morrill W. Collins   Promoted to Senior Captain
Harold Hemingway   Promoted to Senior Captain
Michael A. Bell   Promoted to Captain
Nelson E. Collins   Promoted to Captain
Richard A. Knight   Promoted to Captain

Rookies:

Steven LeTarte	6/13/2000
Dean Frost	6/13/2000
Jason Hyde	8/29/2000

2001   **Officers of Farmington Fire Rescue**

Yrs of Svc.

24	Terry S. Bell, Sr.	Chief
30	S. Clyde Ross	Deputy Chief
23	Timothy A. Hardy	Deputy Chief
41	John O. Bell	Senior Captain
48	Morrill W. Collins	Senior Captain
59	Harold Hemingway	Senior Captain
12	Michael A. Bell	Captain
22	Nelson E. Collins	Captain
22	Richard A. Knight	Captain
17	George L. Barker	Lieutenant
32	Clyde H. Meader	Lieutenant
6	Stanley Wheeler	Chaplain

Firemen:

3	Chet C. Alexander
3	Jonathan M. Alexander
22	Philip R. Allen
23	Melvin L. Bard
19	Peter H. Brennick
15	James A. Brown
23	Stephen M. Bunker
13	Richard A. Chabot
8	Jonathan E. Davis
5	Jason A. Decker
10	David M. Fronk
4	Timothy D. Hardy
13	Rocky Jackson
7	Curtis C. Lawrence
8	Robert E. McCully
17	Eugene L. Mosher
38	Douglas R. Oliver
16	Raymond G. Pillsbury
12	David A Pottle
22	Donald A. Richard
7	Gregory E. Roux
30	Junior E. Turner
19	Randall A. Voter

Additional Information provided by S. Clyde Ross & Terry S. Bell, Sr.

2008

Eugene Mosher	Died	24 yrs. svc.

2009

John "Jack" Bell	Died	49 yrs. svc.
Harold "Stub" Hemingway	Died	67 yrs. svc.

2012

Philip Allen	Retired	33 yrs. svc.

2013

Greg Roux	Retired	19 yrs. svc.

2014

Richard Knight	Retired	33 yrs. svc.
Richard Chabot	Retired	25 yrs. svc.

MEMBERS OF FARMINGTON FIRE RESCUE 2017
CHIEF TERRY S. BELL, SR.
2017 FARMINGTON FIRE RESCUE - YEARS OF SERVICE

OFFICERS:

Chief Terry Bell	40	YRS
Deputy Chief S. Clyde Ross	46	YRS
Deputy Chief Tim Hardy	39	YRS
Captain Mike Bell	28	YRS
Captain T. D. Hardy	20	YRS
Captain Michael Melville	12	YRS
Captain Scott Baxter	12	YRS

FIREMEN:

Doug Oliver	54	YRS
Junior Turner	46	YRS
Stephen Bunker	39	YRS
James Brown	31	YRS
David Fronk	26	YRS
Stan Wheeler, Chaplain	23	YRS
Jon Alexander	19	YRS
Peter Wade	15	YRS
Jon Fortier	13	YRS
James Kiernan	12	YRS
Jennings Pinkham	11	YRS
Teddy Baxter	11	YRS
Patty Cormier	9	YRS
Michael Cote	7	YRS
Brandon Sholan	6	YRS
Kyle Ellis	3	YRS
Joseph Hastings	2	YRS
Christopher Fowle	1	YR
Shawn Latulippe	1	YR
Corey Mills	1	YR
Christopher Nightingale	6	MO
Stanley Cox	6	MO
Sean Zubord	2	MO

Farmington Fire Department Early 1930s
Under Chief Victor Huart

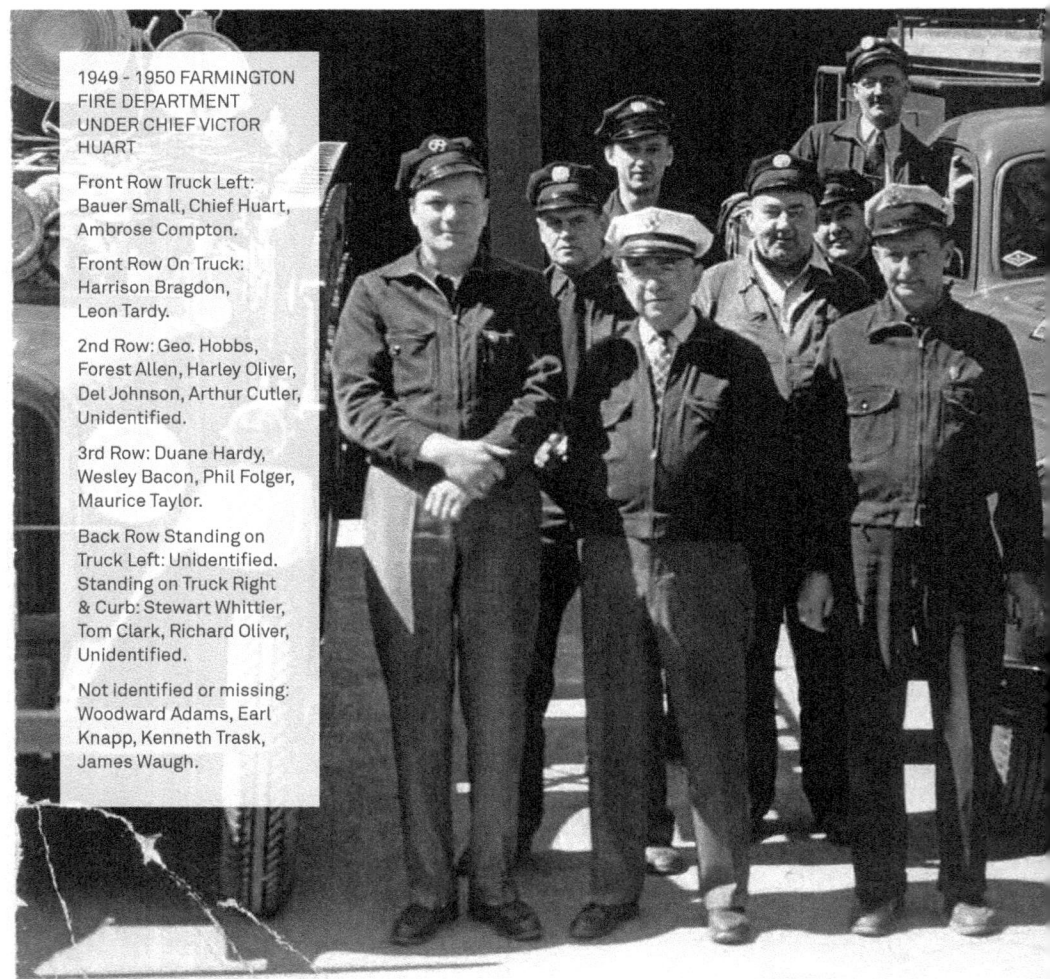

1949 - 1950 FARMINGTON FIRE DEPARTMENT UNDER CHIEF VICTOR HUART

Front Row Truck Left: Bauer Small, Chief Huart, Ambrose Compton.

Front Row On Truck: Harrison Bragdon, Leon Tardy.

2nd Row: Geo. Hobbs, Forest Allen, Harley Oliver, Del Johnson, Arthur Cutler, Unidentified.

3rd Row: Duane Hardy, Wesley Bacon, Phil Folger, Maurice Taylor.

Back Row Standing on Truck Left: Unidentified. Standing on Truck Right & Curb: Stewart Whittier, Tom Clark, Richard Oliver, Unidentified.

Not identified or missing: Woodward Adams, Earl Knapp, Kenneth Trask, James Waugh.

Farmington Fire Department Under Chief Huart 1949 - 1950
1st Row L-R: Stewart Whittier, Richard Oliver, Leon Tardy, Phil Folger, Bauer Small, Chief Victor Huart, Ambrose Compton, Del Johnson, Arthur Cutler.
2nd Row L-R: Unidentified, Tom Clark, Harrison Bragdon, Maurice Taylor, Wesley Bacon Unidentified, Geo. Hobbs, Harley Oliver, Unidentified.
3rd Row L-R: Forest Allen, Duane Hardy.
Not identified or missing: Woodward Adams, Earl Knapp, Kenneth Trask, James Waugh.

FARMINGTON FIRE DEPT. UNDER CHIEF J. BAUER SMALL 1961
(Chief Small in white coat and hat)
Assistant Chiefs L-R: Maurice Taylor & Forest Allen

FARMINGTON FIRE DEPT. UNDER CHIEF FOREST ALLEN 1975
1st Row L-R:Richard Russell, Charlie Grant, Clinton "Kitty" Blaisdell, Harvey Smith, Junior Turner, Art Rogers, Mahlon Moore, Unidentified, Leroy Hazzard.
2nd Row L-R: Jack Bell, Roger Ladd, Stub Hemingway, Tom Cassidy, Guy Brann, Melvin Bard, Geo Hobbs, Chief Allen, Bob McCleery, Harrison Bragdon, Unidentified, Carroll Corbin, Cy Decker.
3rd Row L-R: Robert "Apple" Oliver, Gerald Cookson, David Ferrari, Clyde Meader, Unidentified, Doug Oliver, Pete Durrell, Unidentified, Clyde Ross.

FARMINGTON FIRE DEPT. 1976 UNDER CHIEF ALLEN IN FRONT OF NEW STATION
1st Row L-R: Apple Oliver, David Ferrari, Kitty Blaisdell, Stub Hemingway, Tom Cassidy, Spider Durrell, Maynard Towle, Clyde Ross, Pete Durrell, Terry Warren, Terry Bell.
2nd Row L-R: Gerald Cookson, Chief Forest Allen, Harvey Smith, Richard Russell, Guy Brann, Charlie Grant, Clyde Meader, Geo Hobbs, Roger Ladd, Glen Farmer, Doug Oliver, Jack Bell, Ivan Howard, Howard White, Bob McCleery, Cy Decker.

MEMBERS OF 1977 FIRE DEPT UNDER CHIEF BOB McCLEERY IN FRONT OF THE NEW SQUAD TRUCK

1978 FIRE DEPARTMENT UNDER CHIEF McCLEERY
1st Row L-R: Melvin Bard, Gerald Cookson, Mike Knudtson, Tom Cassidy, Lt. Clinton Blaisdell, Spider Durrell, Charles Grant, Ivan Howard, Terry Bell.
2nd Row L-R: Harvey Smith, Asst. Chief Clyde Ross, Chief Bob McCleery, Dep. Chief Glen Farmer, Asst. Chief Richard Russell, Asst. Chief Norm Collins, Capt. Jack Bell, Bradford Moore.
3rd Row L-R: Tim Hardy, Lt. Clyde Meader, Doug Oliver, Roy Hazzard, Morrill Collins, Roger Ladd, Howard White, Carroll Corbin.
4th Row L-R: Junior Turner, Apple Oliver, David Ferrari, Mahlon Moore, Pete Durrell.

1983 FIRE DEPARTMENT

1985 FIRE DEPARTMENT
1st Row L-R: Lt. Clyde Meader, Greg Wilson.
2nd Row: Peter Brennick, Tom Cassidy, Lt. Kenneth "Spider" Durrell, Junior Turner,
Ray Pillsbury, Geo. Barker, Carroll Corbin.
3rd Row: Art Rodgers, Eugene Mosher, Don Richards, Asst. Chief Clyde Ross, Chief Robert McCleery,
Deputy Chief Glen Farmer, Asst. Chief Terry Bell, Asst. Chief Norm Collins, Ivan Howard, Jim Kiernan.
Back Row: Lt. Morrill Collins, Robert 'Apple" Oliver, Steve Bryant, Roger Ladd, Doug Oliver, Lt. Tim Hardy,
Melvin Bard, Randy Voter, Capt. John "Jack" Bell, Harold "Stub"Hemingway, Brad Moore, Harvey Smith,
Phil Allen, Lt. Nelson Collins, Mike Kiernan, Richard Knight.

1992 FIRE DEPARTMENT
1st Row L-R: Dick Knight, TD Hardy, Geo. Barker, Junior Turner, Rocky Jackson.
2nd Row: Greg Roux, Phil Allen, Tim Hardy, Clyde Ross, Chief McCleery, Terry Bell, Ray Pillsbury.
3rd Row: Jon Davis, Morrill Collins, Mel Bard, Bobby McCully, Randy Voter, Stub Hemingway, Dick Chabot,
Clyde Meader, Eugene Mosher. 4th Row: Steve Bunker, Doug Oliver, Jack Bell, Mike Bell, Jim Brown,
David Fronk, Nelson Collins, Curtis Lawrence, Jason Decker.

1999 FARMINGTON FIRE DEPARTMENT UNDER CHIEF ROBERT McCLEERY
1st Row L-R: Capt. Jack Bell, Deputy Chief Clyde Ross, Chief Robert McCleery,
Asst. Chief Tim Hardy, Deputy Chief Terry Bell, Capt. Harold "Stub" Hemingway.
2nd Row: Richard Chabot, Lt. George Barker, David Pottle, Junior Turner, Robert McCully,
Lt. Richard Knight, Don Richards, Jon Alexander, TD Hardy.
3rd Row: Lt. Mike Bell, Phil Allen, Melvin Bard, Jason Decker, Jim Brown, Lt. Clyde Meader,
Chet Alexander, Jon Davis, Steve Bunker.
4th Row: Doug Oliver, David Fronk.
Not present: Capt. Morrill Collins, Lt. Peter Brennick,
Lt. Nelson Collins, Chaplain Stanley Wheeler

2004 FIRE DEPARTMENT UNDER CHIEF TERRY BELL
1st Row L-R: Capt. Mike Bell, Capt. Richard Knight, Dep. Chief, S. Clyde Ross, Chief Terry Bell,
Dep. Chief Tim Harcy, Capt. Stub Hemingway, Capt. Jack Bell.
2nd Row: Jon Paul Fortier, Abbie Davenport, Richard Chabot, Robert McCully, Jon Alexander, Tim D. Hardy,
Jason Decker, Jon Davis, Robert Mosher, Randy Voter, Greg Roux, Glen Speight, Junior Turner.
3rd Row: Phil Allen, Steve Decker, Melvin Bard, Anthony Russo, Peter Ward, David Fronk, Eugene Mosher,
James Brown, Stanley Wheeler, Steve Bunker.

2013 FIRE DEPARTMENT UNDER CHIEF TERRY BELL
1st Row L-R: Stephen Almquist, Junior Turner, Doug Oliver, Dep. Chief S. Clyde Ross, Chief Terry Bell,
Dep. Chief Tim A. Hardy, Mike Cote, Patty Cormier.
2nd Row: Brandon Sholan, Greg Roux, Lt. Tim D. Hardy, Teddy Baxter, Eric Gilbret, Capt. Mike Bell, David
Fronk, Peter Wade, Chaplain Stan Wheeler, Capt. Richard Knight, Jennings Pinkham, Lt. Mike Melville,
Tyler Poul'n, Andy Cote, Phil Allen, Richard Chabot.

Chapter 6

FATAL FIRES 1914 – 1999

"When you get inside and make a good save, that's a good feeling, but that's not how it always plays out. The hardest thing for any firefighter to deal with is if there people inside, people you can't rescue, or for whom it's already too late, especially children. It will take the life right out of the most hardened person."

~ Chief Bob McCleery

1914	July 29	Miss Adele Rossman, German maid for Dr. Virgil Coblentz, Brooklyn, NY, died at NW Shore of Clearwater, Industry (More recently the Sherwood Swain camp) in kitchen fire caused by explosion of a kerosene can; Assisted Allens Mills.
1936	April 1	Mrs. Clement Luce and three children died in fire at the junction of Lower Main Street and High Street.
1964	January 15	Albert Black, North Main Street, 4 year old son died in fire.
1970	November 19	Lester C. Hutchinson died in fire that destroyed the house owned by Howard Pease in back of the B & M Corn Shop by the Center Bridge.
1974	January 26	Bernard Allen's son died in Chesterville fire that destroyed trailer; Assisted Chesterville.
1976	April 15	Jean (mother) and Troy Hunter (2 year-old son) died in fire at 79 North Main Street.
1979	December 18	Doug Swain house, East Wilton, John Pollard died in fire; Assisted Wilton.
1982	April 5	Joan Colleen Hickey (Age 45) and daughter Amy Bicknell (Age 16), High Street and corner of Lake Avenue, died of smoke inhalation.
1982	September 20	Elisha Mae Tyler (2 years old), Route 4 Strong Rd, died from smoke inhalation in a fire that destroyed her home.
1984	May 2	Leroy and Mildred Hammond, Hammond Road off Route 133 East Wilton, died in early morning fire; Assisted Wilton.
1990	March 6	Susan Morse home Davenport Hill, Jay; 2 Fatalities; Assisted Jay.

"Good common sense is still the best method of fire prevention. I cannot urge the citizens of Farmington enough in the proper installation, care, and maintenance of their chimneys and wood burning stoves. Have them inspected and cleaned yearly. I strongly recommend the use of some type of warning device, either heat or smoke alarm. Test these devices yearly, and replace the batteries immediately if they chirp. Carbon monoxide monitors should be installed as well. Be careful with the disposal of ashes. Do not place them in plastic pails, bags, or cardboard boxes, or dump them beside a building. Know where your children and family members are. It could save their lives."

~ Chief Bob McCleery

Chapter 7

FIRES OF $50,000 PLUS 1850 - 2000

1850		Fire destroyed the square between Main Street, Broadway, and Exchange Street.
1886	October 24 -	$300,000 Fire destroys Farmington-Main, Academy, South and Pleasant Sts.
1932	March 31 -	Franklin Farmers Corn Factory, Intervale Rd. destroyed by fire.
1932	April 30 -	Fire damages Music Hall on Broadway.
1933	February 2 -	Fire destroys New Sharon School.
1935	September 19 -	Morton Motor Company garage, Main Street, destroyed by fire.
1937	March 26 -	Post Office, West Farmington badly damaged.
1938	March 20 -	First Baptist Church, Academy Street, destroyed.
1939	November 16-	Music Hall, A&P Store, Coffee Shop, Hardy's Pharmacy and Edith Hardy apartment seriously damaged.
1940	June 27 -	Besson Farm, North Chesterville, destroyed by fire.
1940	August 5 -	Linwood York Farm, Titcomb Hill, destroyed.
1941	November 19 -	Gordon's Mill, Farmington Falls, destroyed.
1944	May 12 -	Town Farm, Town Farm Road, House, barn, and 19 head of cattle lost.
1944	May 22 -	Herman Lidstone Barn, ell and part of house now (Northern Lights), Routes 2& 4.
1945	February 26 -	Normal School Annex, High Street, destroyed.
1946	February 12 -	Linwood York, Titcomb Hill, Barn and cattle lost.
1946	July 4 -	F. O. Smith Mill, New Vineyard, destroyed.
1946	August 12 -	Fair Grounds, north cattle sheds, destroyed.
1947	July 12 -	Metcalf Mill, West Farmington, wood turning mill, destroyed.
1947	August 4 -	Carson's Laundry, North Main Street, destroyed.
1947	August 12 -	Brackley's Mill, Strong, destroyed.
1948	February 18 -	Pearson's Barn, Ernest Doyen house, barn, 36 head of cattle lost, and Kenneth French house destroyed.
1949	June 13 -	New Sharon High School, destroyed.
1951	December 15 -	C. W. Steele Company (Barrows Block), Main St., two buildings destroyed.
1954	March 31 -	Arthur Gagne's Store, New Vineyard, destroyed.
1954	October 28 -	Frank Osborne Farm, Woodcock Hill, destroyed.
1960	August 28 -	Victor Armstrong, Strong Rd, chicken barn destroyed.

1960	November 12 -	Medomak Canning Co., New Sharon, corn factory and warehouse destroyed.
1961	February 4 -	John Newcomb, Wilton Road, house, barn and Ray-Jon Kennels destroyed.
1961	December 3 -	Gladys Bailey, Bailey Hill, barn and contents destroyed, house saved.
1964	November 15 -	Town of Phillips, Upper Village, several businesses in the Upper Village lost.
1965	January 1 -	Bouffard's Furniture Store, Main St., heavy damage.
1965	April 3 -	Newman Motors, Broadway, destroyed.
1965	October 25 -	Brackley's Mill Strong, destroyed.
1968	February 2 -	Fair Grounds, New Horse Barn, several horses lost.
1968	April 9 -	Fair Grounds, New horse stables, back of grandstand destroyed.
1968	May 16 -	Merritt Averill buildings, Starling St., West Farmington, destroyed.
1969	December 6 -	Town of Strong, westerly side of Main St., several buildings lost.
1970	June 11 -	Harold Hardy, Perham Hill, barn struck by lightning and destroyed.
1970	July 2 -	Clint Conant, back road to Strong, barn destroyed by lightning.
1971	January 4 -	Barker's paint shop and storage garage, High Street, destroyed.
1971	April 10 -	Beal Block, Main St., Phillips, all but 2 buildings in the business district destroyed, 10 businesses and 4 family homes.
1971	July 10 -	Wilton business block in center of town, destroyed.
1973	August 28 -	West Mills Church, struck by lightning, destroyed.
1975	April 2 -	Drummond Hall Block, Broadway, destroyed. 3 businesses lost.
1979	December 1 -	Smith's Mill, New Vineyard, destroyed.
1980	May 10 -	Wilton Academy, destroyed by fire.
1983	August 17 -	Starbird Lumber (United Timber), Falls Road, destroyed.
1983	December 17 -	Country Haven Nursing Home (on corner), destroyed.
1984	February 28 -	Robert Fenn, North Chesterville, Barn & sheep lost.
1985	February 10 -	Golden Dragon Restaurant, Mt. Blue Shopping Center.
1985	March 21 -	Dorman Block, Main Street, Wilton, destroyed.
1986	March 9 -	Sweetser Mill, Route 43 Temple Rd., destroyed.
1986	September 14 -	Richard Davis New Sharon, Barn destroyed.
1988	March 12 -	Tom Williams Apartment House, 120-121 Main Street.
1989	December 12 -	Jordan, Roux, Egars Apartment Building, 40-42 Broadway, destroyed.
1990	February 2 -	New Attitude Clothing (D. Pike), Wilton Rd., destroyed.

1992	January 19 –	Apartment House, 1 1 Quebec Street.
1992	October 10 –	Steve' s Market, Dryden, destroyed.
1993	March 22 –	Country Manor Restaurant, Ronald Rackliff, Falls Road.
1993	May 17 –	Ferrari Clothing Store, Corner of Broadway & Main, $350,000 loss.
1993	October 26 –	Inn Town Lounge, Jay.
1994	January 20 –	Beef Barn, Bean' s Corner, Jay, destroyed.
1994	May 22 –	Apartment House, 79 North Main Street, destroyed.
1994	September 20 –	D & D Ridley, Farmington Falls-New Sharon, Mason Road, destroyed.
1994	October 30 –	Hosmer-Edwards American Legion Post, Wilton.
1995	July 1–	Sandy River Farm, L.Herbert York, Lost barn and house heavily damaged. Lightning.
2000	January 4 –	Robert Austin Apartments, 5 Thomas McClellan Road, West Farmington, destroyed.

Chapter 8

FIRE & RESCUE 1885 - 2000

| 1885 | 6/03 | William A. Titcomb barn; Lightning strike; Total loss |
| | 7/27 | William Stoddard house & barn; Accidental; Partial loss |

| 1896 | 10/17 | Mrs. Marion Gray house & barn (Lyman F. Fales); Total loss |

1897	2/04	S. R. & J. A. James house; Total loss
	7/10	E. T. Abbott house & barn; Total loss
	12/24	S. G. Preston house & barn; Chimney fire; Total loss

1898	4/22	G. W. Ranger Saw Mill & Grist Mill; Fire started in Mill; Total loss
	4/22	G. T. Currier Store; Caused by Mill fire; Total loss
	8/24	H. B. Searles house; Partial loss
	9/12	Nathan G. Whittier house, barn & carriage house; Total loss
	10/05	U. S. Gardiner House & barn; Total loss

| 1899 | 2/12 | Isace Russell house; Chimney fire; Partial loss |

1900	1/10	Hinckley Clothing Co.; Store & house; Partial loss
	3/29	Andrew Norton house; Total loss
	5/04	Harvey W. Lowell house; Total loss
	6/26	Charles Eaton farm & buildings; Chimney fire; Total loss
	7/12	Martha A. Dunsmore house; Total loss
	7/12	Albert Gilbert house; Total loss
	7/16	Samuel Tarbox stable & contents; Lightning strike; Total loss
	7/21	G. A. Abbott dwelling & boarding house; Cause Unknown; Total loss
	7/26	A. H. Abbott library, boys school & grounds; Cause unknown; Total loss

| 1901 | 12/06 | Russell Bros & Estes Co. Factory; Cause lamp; Partial loss |

1902	5/13	A. E. Scribner dwelling; Defective chimney; Total loss
	5/13	John Briggs house (Frank Ames); Scribner sparks fire; Total loss
	5/13	Granville Hackett house (Howard Parker occup.); Scribner sparks; Total loss
	5/13	A. F. Gammon (Arthur Farmer) Mill; Scribner sparks; Total loss
	5/18	Charles Plaisted dwelling; Chimney fire; Total loss
	5/18	L. T. & F. L. Childs dwelling; Sparks from Plaisted fire; Total loss

1903	4/21	S. Knowlton & Son Store; Partial loss
	5/01	N. D. Keith dwelling; Chimney fire; Total loss
	5/01	E. S. Bragg dwelling; Sparks from Keith fire; Total loss
	5/01	F. N. Wilder Saw Mill; Sparks from Keith fire; Total loss
	5/01	Henry Pike dwelling; Chimney fire; Total loss
	5/15	A. H. Dyer dwelling; Sparks from chimney fire; Total loss
	5/28	E. A. Pinkham Carriage Shop; Lantern broke; Total loss
	7/16	J. A. Jones barn; Cause unknown; Total loss
	12/06	Mrs. E. N. Allen dwelling; Defective chimney; Total loss

1904	11/06	C. W. Cameron dwelling; Defective chimney; Total loss
1907	7/11	J. L. Stowe dwelling; Defective chimney; Total loss $2500
	7/20	Charlton Furbush barn & stable; Lightning strike; Total loss $1200
1908	2/14	Town of Farmington Schoolhouse; Cause Unknown; Total loss $700
1909	4/17	E. H. Marwick (C. F. Davis) Store; Cause Unknown; Total loss $3400
1912	2/18	James E. McKeown dwelling; Cause Unknown; Partial loss $1300
1914	7/29	Dr. Virgil Coblentz (NY) Clearwater cottage; Kerosene can explosion; Adele Rossman, young German maid died; Assisted Allens Mills; Bldg. loss $6000. Young woman died in kitchen fire caused by explosion of a can of kerosene while pouring oil into cook stove to increase fire, blaze was communicated to contents of can and explosion took place. Cottage destroyed.
1919	2/28	Russell Bros. & Estes Co. Birch Mill; Cause Boiler; Total loss
	5/24	G.L. Howes & D.L. Wyman dwelling; Defective chimney; Total loss $2500
1922	2/19	J. C. Dustin dwelling; Defective chimney; Total loss
	6/01	H. A. Crowell farm; Defective chimney; Total loss
	7/13	Masonic Building & Frank Howatt Barber Shop: Partial loss $2500
1923	2/19	Florence B. Clark dwelling; Stove pipe fell out; Partial loss $2500
	3/30	Lillian Paine Store (Dolbier & Hardy); Defective chimney; Partial loss
	7/21	Dolly Smith dwelling; Cause unknown; Total loss $1000
	9/19	A. S. Lowell dwelling; Kerosene stove; Partial loss
	9/29	Elbridge Luce stable & garage; Cause unknown; Total Loss
	10/03	T. T. Gordon dwelling & barn; Cause unknown; Total loss $3000
	12/17	Charles H. Pierce dwelling; Furnace; Partial loss
1924	6/01	Arthur Gaskell dwelling & barn (Archie Pratt); Unknown; Total loss $4000
	6/01	Ralph Parker dwelling & barn; Sparks from Pratt fire; Total loss $4000
	6/23	Forest Smith Dance Pavilion; Cause unknown; Total loss
	9/15	C. A. Pinkham Carriage Co.; Sparks from chimney; Total loss $5000
	9/17	Linwood York dwelling & barn; Cause unknown; Total loss $5000
	11/09	Lillian M. Horn hay storage (A. D. Horn Est.); Causes unknown; Total loss
1925	1/10	H. S. Campbell dwelling; Defective wiring; Total loss $7000
	6/14	Otis B. Littlefield dwelling; Electric flat iron; Total loss $3500
	7/30	Warren Smith dwelling; Defective chimney; Total loss
	9/12	Ernest Vining dwelling & barn; Lightning strike; Total loss $3200
	10/17	M. W. Petrie barn; Total loss
	11/26	Marcus H. Mitchell barn; Total loss
1926	1/18	Mrs. Thomas Davis dwelling; Hot coals falling on carpet; Partial loss
	6/05	Elizabeth Dustin dwelling; Defective chimney; Total loss
	12/19	Albert Allen camp (M. Mitchell); Sparks from chimney; Total loss

1928 Fire at Wilford McLeary Store
Site of former J.J. Newberry's and current Liquid Sunshine Boutique
165 Main Street
$2500 Partial Loss
Photo Courtesy of Don DeRoche

1927 5/04 Hannibal Russell dwelling; Cause unknown; Total loss
9/25 Clarence Hewey Garage; Cause unknown; Total loss $2000

1928 1/11 Alfie McLeary (Wilfred McLeary Store); Wiring; Partial loss $2500
4/17 F. G. Colburn (F.C.A.S) Stable; Sparks from Stove; Total loss $1500
8/22 Ellery Farmer (A. S Lowell); Partial loss
12/4 Hiram Nichols Dwelling; Chimney; Total loss $5500
12/31 Amos Cox Dwelling (McLeary Est.); Unknown; Total loss $2000

1929 7/11 Lutie Holley Dwelling (Geo. Holley); Oil stove; Total Loss $3000
10/02 Geo. Wade & Richards Garage/House; Wiring; Loss $2500
10/02 Farmington Falls Electric Co., Power Station & Mill; Sparks from garage fire; Total loss $5000
10/02 Elvier Collins Dwelling; Sparks from garage fire; Loss $1500
10/02 Longa Whitcher Dwelling; Sparks from garage; Total loss $3500
10/02 Adelbert Williams Dwelling (Lewis Oliver); Sparks from garage; Total loss $3000

1930	7/19	Frank Hemingway Dwelling; Wiring; Partial loss $2000
	7/20	Dr. E. E. Russell; Stable & House; Spontaneous Combustion; Partial loss $3000
	10/23	D. T. Williams Dwelling & Barn; Unknown cause - started in barn; Total loss $5000
1931	3/19	C. J. Macomber & E. J. Rathey; Wood Turning Mill & Garage; Cause Electric Motor; Partial $2500
	9/19	Chas. Wyman; Fairbanks; Barn & shed
	10/13	Carroll Whitney; Fairbanks; Store
	10/20	Stoddard House; Broadway in Town
	11/01	John Norton; Perham Street
1932	11/01	Horn Stables; Occup. Newman Motors; Maxim pumped 5 Hrs. & 20 Mins.
	1/31	Franklin Farms Corn Factory; 8 Hrs.; 3500 ft. of hose laid
	2/19	Guy Sweetser; Lincoln Street
	4/30	Music Hall
	9/03	Joe Nichols Garage; Farmington Falls
	9/09	Strong; 4 Hrs.; No report
	10/31	Clinton Savage; Sawyer barn; Farmington Falls
	10/27	Arthur Rackliff house; Allens Mills
1933	1/29	Kingfield
	2/16	High School; New Sharon
	4/13	Tibbetts farm; Falls Road
	5/02	F. N. Blanchard barn; Wilton; Total loss
	8/06	Holly Farm
1934	6/26	Ralph Carrier farm; Chesterville
	10/06	Raymond Hiltz farm; North Chesterville Road
1935	5/10	Hardy Farm; Farmington Falls
	9/19	Morton Motor Co. Garage; Main Street; 9 Hrs.; Totaled (Farmington Fair Week)
	12/19	Charles Davis house; Falls Road
	12/30	Leonard Keith mill; North Chesterville; Total loss
1936	4/01	Clement Luce; Lower Main Street; Mrs. Luce and 3 children died.
	9/26	Lamkin Farm; North Chesterville, Total loss
	10/19	Edna Robin; Lincoln Street
	11/08	Grounder House; Knowlton Corner Road
	11/10	Pietpis Farm; North Chesterville
	12/30	L. W. Harris store; Maiden Lane; Harness Shop - Leather Shop
1937	3/02	Elmer Rathey house; West Farmington
	3/10	Ed Rackliff farm; Back Falls Road; Stable and ell
	3/15	Maud Norton; Church Street; 2nd Floor Rent
	3/26	Post Office; West Farmington
	10/08	A. A. Nickerson house; Wilton Road; Fire around chimney
	12/01	North Church; Basement

1938	3/20	First Baptist Church; Academy Street; Total loss
	6/28	Clarence Hewey; Perham Hill; Total loss
	7/09	Bert Hardy farm; Red Schoolhouse Road; Twin barns; Total loss
	10/06	Simon Luce house; Depot Street; Roof damage from fire
	11/19	Clark Mill; West New Portland; Total loss
	11/27	North Church; Basement
1939	3/03	Arthur Gaskell house
	3/31	Leonard Doble house; Allens Mills; Total loss
	8/13	Kempton Lumber Co.; Rangeley; 1 Hr. 10 Mins. water put on fire; Total loss $50,000
	9/26	Mrs. Simon Luce house; Roof fire
	10/07	Metcalf's Mill; West Farmington; Boiler house
	11/16	Music Hall, A&F, coffee shop, Hardy's Drug Store, Edith Hardy Apartment; Major loss
	12/05	Pinkham's garage; Keyes Block (parking lot today); Major damage
1940	2/13	George Fredericks; New Vineyard Road; Garage; Total loss
	2/17	John Jones; Perham Street; Major damage
	6/27	Besson farm; North Chesterville; Barn, shed and part of house destroyed
	8/05	Linwood York farm; Titcomb Hill; Barn, shed & ell; Total loss
	8/23	Brown's Jewelry Store; Broadway; Basement; Minor damage
	9/18	Leon Brown; Farmington; Ell and house; $4500 loss
	10/09	Salle farm; Voter Hill; Total loss
1941	4/12	Bill Neal; Suburban District; Garage; Total loss
	6/09	Frary's Mill; Square yard and woods; High winds
	7/25	Manning Titcomb farm; Barn struck by lightning; Total loss
	8/11	C. W. Hicks farm; Knowlton Corner Road; $2000 loss
	9/18	Leon Hall house; Lower Main Street; Caused by fireworks at Fairgrounds
	10/22	Kenneth Jellison house (owned by B. Butler); Perkins Street; Major damage
	11/10	Gordon's Mill; Farmington Falls; Pumped 6 ½ hrs. from river; 1350 ft. hose; Total loss
	12/11	Baptist Church; Rangeley; Fast run; Under control; Stand by
	12/22	Eddie Tague house; Near Center Bridge; Total loss
1942	12/14	Bailey house; Perkins Street; Damage from chimney fire
1943	1/05	Harold Hall; North Chesterville; Fire in outhouse; Cause hot ashes
	11/04	Howard Fuller home; Maple Ave; Barn and ell
	12/25	Pete Blanchard; Chesterville; Buildings; Total loss
1944	4/01	Heziciah Mason; Backus Corner; Out buildings, hen house; Saved hens
	4/28	Byron Small house; Roof fire; Extensive damage
	5/12	Town Farm; Barn and house; 19 head of cattle lost; $3500 loss
	5/22	Herman Lidstone; Wilton Rd; Barn, ell and part of house; Tractor backfired; $6000 loss
	6/27	Zimmerman farm (now Davis farm); New Sharon; House; Total loss
	8/11	Howard Pease wood lot; 2 acres
	8/17	Owen Mann farm; Porter Hill; Lightning; No water; Had to let burn
	10/24	Maude Norton house; Church Street; Livingroom; Short circuit wiring; $1000 loss.

1945	2/15	Pearson farm (Del Paradis house); Falls Road; House lost; Barn saved
	2/26	Normal School Annex; High Street; Total Loss; $12,000
	3/19	Tom Kelley house; Farmington Falls; Total loss
	4/11	Arnold Porter house; Farmington Falls; Couldn't reach river; Total loss
	7/27	John Gilkey's monument shop; Interior ablaze; $2700 loss
	10/20	Lawrence Smith; Middle Street; Cellar fire; $650 loss
	11/21	Perley Brown house; Below Fairgrounds; 2nd Call 1 Hr. later; Total loss
1946	1/03	Stanley Robbins farm; North Chesterville; Lost house; Saved barn
	1/27	Bert Knowles home; Temple Road; Total loss
	2/12	Linwood York farm; Titcomb Hill; No water; Barn and contents; Total loss
	4/07	Lester Sprague house; Fairbanks; House and ell gutted
	6/14	Gus Sawyer farm; Back Temple Road; Burned flat; No water
	7/04	Maurice Dill clothing store: Fire cracker in awning; Roof and front of 2nd Story; $2500 loss
	7/04	F. O. Smith Mill; New Vineyard; Pumped 1 ½ Hrs.; Broke 9 ft. long hose; Total loss
	7/06	Manter house; Maple Ave; Barn; Total loss; $3500
	8/12	Fairgrounds; North cattle sheds lost; Saved one south shed and ½ other; $6500 loss
	12/16	Norman Weymouth; Pleasant Street; Heavy damage
1947	7/02	Metcalf Mill; West Farmington; Wood Turning Mill; Total loss
	7/06	Scales farm; Temple; Lightning strike; Buildings; Total loss
	7/17	Leonard Brooks; West Farmington; Barn Total loss; Ell damaged
	8/04	Carson's laundry; North Main Street; Total loss. Smith's filling station (south side of laundry) and Jalbert house (north side); Badly scorched
	10/12	Brackley's Mill; Strong; Assisted Kingfield & New Vineyard; Total loss $50K
	10/21	Joseph Gilbert farm; Perham Hill; Total loss; No water
	10/23	Elmer Robbins farm; Allens Mills Road; Barn and cattle lost; 15 head cattle
	10/25	Lawrence Boyce house; Temple; House, ell and barn; Total loss; No water
	12/04	Forster Mill; Strong; Assisted Wilton, Phillips, Kingfield & Strong; Pumped 4 Hrs. from Valley Brook; 20 lengths 2 ½ in. laid; Mill Total loss
	12/09	Harold Beal; Kings Corner; Barn and house; Total loss; No water
1948	2/10	L. Kenniston farm; Industry; House lost; Barn saved; Neighbor hauled water
	2/18	Pearson Barn; Fall Road; Destroyed; High winds blew sparks into cupola of Earnest Doyen barn; Destroyed barn, house and 36 head of cattle lost; Then to Kenneth French house; Total loss $50,000
	4/19	Herbert Lakin farm; Barn lost; House saved
	4/23	Vienna store; Assisted Mt. Vernon and Readfield; Store lost
	8/06	Heziciah Mason garage; Holley Road; Total loss $2000
	8/27	Roger Young house; Maple Ave; $2500 loss
	9/14	Old Clark farm; Upper Perham Street; House lost; Barn saved; $5000 loss
	12/02	Orville Meisner house; Farmington Falls; House ablaze; $2000 loss

1949	3/31	Logan Luce farm (Formerly Tufts); Barn and 16 head of cattle lost; House saved; $15,000 loss.
	6/13	New Sharon School House; Total loss; $60,000
	11/16	Stoddard House; Broadway; Rear damage; $2500 loss
	12/01	Phillip Ladd house (Jake Ladd); Strong Road; House heavily damaged
	12/04	Frank Dallasandro house; Anson Street; 2nd Floor damage; $4500 loss
	12/09	Phillip Luger farm; Red School House Road, Wilton; House destroyed
1950	2/20	David Holmes; Strong Road; 1½ story house; $2000 loss; High wind/18 deg.
	3/02	Lloyd Pratt house; Chesterville; Badly gutted; Assisted by New Sharon
	9/24	Roland Farrington farm; North Chesterville Road; Barn & house; 1 Bull and 2 Heifers lost
	12/15	Jerome Holley farm; Temple Road; House badly damaged; $3000 loss
1951	2/02	Norman Adams house; West Farmington; Living room and upstairs bedroom; $3500 loss
	7/27	Waldron Marble barn; North Chesterville; House saved, Barn & contents lost; $8000 loss
	8/20	Lawrence L. Smith barn; Allens Mills Road; Total loss $1500
	10/25	John Fox farm (Bert Knowles); Temple Road; Barn and house lost; $6000 loss
	12/15	C. W. Steele Co.; Barrows Block; 2 wooden buildings; Total loss $30,000
	12/31	Charles Hardy house; New Vineyard Road by Barker Brook; Shed, ell and kitchen destroyed

By Staff Photographer

FIRE IN FARMINGTON—Farmington and West Farmington firemen fight a $30,000 fire Saturday that destroyed two buildings on Farmington's Main Street. Damaged were the C. W. Steele Company, Press Herald News Bureau and Oscar Beedy barber shop. Two families were driven from their apartments. Firemen battled the blaze four hours.

2 Farmington Blocks Burn; Loss $30,000

(Special Dispatch)

Farmington, Dec. 15.—A $30,000 fire swept two buildings in the Main Street business district here today and threatened several other structures.

The blaze started in a two and a half story wooden building occupied by the C. W. Steele Company, oil burner concern and spread to an adjacent building housing the Press Herald News Bureau and a barber shop.

Both buildings are owned by Clyde I. Barrows.

Four Rescued

Mr. and Mrs. James E. Pratt and two children, occupants of an apartment over the Steele Company, were led to safety through dense smoke by Clifton Quimby of Auburn and Joseph Senko of Lewiston, truck drivers. Pratt, a navy enlisted man, was home on week-end leave.

The fire was discovered by Quimby and Senko as they were eating in Stowell's Restaurant in an adjoining building. They warned Mr. and Mrs. Ellsworth Decker and two sons, who occupied an upstairs apartment, and then went to aid the Pratts.

Meanwhile the blaze spread to a one-story structure housing the News Bureau, and Oscar Beedy's barber shop. Press Herald Correspondent Edward T. Maguire

Farmington Fire:

(Continued from Page One)

also operated a variety store in the bureau's quarters.

Smoke Damage

Stowell's Restaurant and the Decker apartment were heavily damaged by smoke and water, and a one-story building housing Bonney's Lunch was also damaged by smoke.

Firemen working in 10-degree weather halted the flames before they reached the Western Union building and other business establishments.

Although a northeast wind whipped the blaze at times six inches of fresh snow blanketed rooftops and prevented new fires from starting. The damage was estimated by Fire Chief J. Bauer Small, who said he believes the fire was caused by a torch being used to thaw pipes.

1951 Fire at Main Street Business District
C. W. Steele Fuel Company, Press Herald News Bureau
& Oscar Beedy Barber Shop
Property Owned by Clyde I. Barrows
Estimated $30,000 Loss

Photo Courtesy of Don DeRoche and Portland Press Herald

1953	3/21	Kenneth Ladd trailer home; Lincoln Street; Destroyed; $2000 loss
	4/07	Henry Metcalf house West Farmington Route 43; Heavy damage to ell, house and shed
	5/07	McLeary Hardware Store; Broadway; Rubbish; Spread to Drummond Hall attic; $3500 loss
	7/12	Ed Welch house; Fairbanks; Total loss
	11/25	Jack Whelpey house; West Farmington; Interior badly damaged; $3000 loss
1954	1/04	Joe Philbrick house Farmington Falls; Ell destroyed, house badly damaged; $5000 loss
	1/14	Buddy Foster house; Farmington Falls Road; Heavy damage to interior; Used shotgun to blow hole in floor
	2/21	Former Dr. Ross house; Pleasant Street; Northeast News occupied 1st Floor; Bill Loring occupied 2nd Floor; Barn, ell, part of house lost; $18,000 loss
	3/22	George Ladd house; New Vineyard; Barn, ell and 2nd Floor of house; Heavy damage
	3/31	Arthur Gagne's Store (Fred Luce); New Vineyard; Assisted New Vineyard and Strong; $20,000 loss
	5/03	Murdock Petrie; Lower High Street; Barn, shed and ell destroyed
	10/14	Floyd Clements farm; Chesterville Hill; Barn destroyed
	10/18	Parker Springs Restaurant; Wilton Rd; Heavy damage to kitchen & building
	10/28	Frank Osborne farm; Bailey Hill; Smoke and heat damage to house and ell; Barn, milking equipment and 50 - 70 tons of hay lost; Pumped water from Pond in field; No insurance
1955	2/01	Robert Perkins; Holley Road; Lost house; Barn and ell saved; $8000 loss
	3/08	Clyde Richardson farm; Temple Road; Barn total loss
	4/06	Irene Tuttle house (owned by Harold Hardy); House destroyed
	5/09	Bert Moore woods; Holley Road; 10 acres; 4 hours
	7/17	Granville Storer; Chesterville; House lost
	9/23	Stoddard House; Damage to building from heat and smoke; $2500 loss
	12/30	Earl Luce farm; Holley Road; Roof and attic caused by chimney fire
1956	7/25	Macomber Mill; West Farmington; Birch mill lost; Main mill saved; $10,000 loss
	10/27	Kenneth Ladd house; Strong Road; Total loss; Nothing saved
	12/03	Macomber Mill; West Farmington; Birch mill; $1500 loss
	12/31	Lawrence Smith house; Middle Street; House total loss
1957	3/19	Wesley Mitchell farm; Back Strong Road; All buildings lost
	4/03	Sherman Tracy farm; New Sharon; House saved; Barn lost
	4/05	William Downs house; New Vineyard; House and ell lost; Barn saved
	6/25	Orville Gardner house; Farmington Falls Rd; House & contents; Total loss
	12/16	Raymond Tyler house; New Vineyard Road; 1 ½ Story; Total loss;1:15 AM - 3:50 AM; Return call 5:15 AM - 7:45 AM; lost shed & rest house

1950's - 1960's
Photo Upper Left: 1965 Newman Motors Fire
Photo Lower Right: 1951 C. W. Steele, Barrows Block Fire

Photos Courtesy Town of Farmington

1958	1/04	E. Robbins farm; Industry; House and barn; Total loss; Burned flat
	2/04	Moon Glow Motel; Wilton Rd; Store room and laundry gutted; Smoke and water damage; Caused by blow torch
	3/15	Russell McKinnon house; Federal Row; House and antiques lost; Walked to fire, road not plowed of deep snow; No water; $30,000 loss
	3/26	Bert Trask house; West Farmington; Ell, shed and part of house; Total loss
	5/15	Jordan Tarbox house; Corner of South & Main St; Apartment of Rose Lamberton; Damage to apartment only
	5/16	Neil farm; Chesterville; Chicken house destroyed; No water
	5/26	Pillsbury house; North Chesterville Road; No one home; Total loss
1959	2/01	Ralph Wood house; Strong Road; Severe damage to house; Returned to house later in day for fire in basement
	2/03	Ralph Wood house; Strong Road; Completely in flames; Nothing could be done; Let it burn and went back to station
	10/05	James Sutliffe house; Heavy smoke damage from flooded oil burner
	12/27	George Frary house; Anson Street; Minor damage; Defective wiring
	12/30	Roland Farrington house; Wilton Rd; 2nd Floor heavy damage

1960 Fire at Medomak Canning Company , New Sharon
Estimated $2 Million Loss

Photo Courtesy of Lewiston Sun Journal

1960	3/06	Pettingill house; North Chesterville; Total loss; No one living there
	3/26	Albert Kilgore; Temple; House total loss; No one home
	8/28	Victor Armstrong chicken farm; Strong Road; Barn and contents lost; House saved; 7 hour fire

	11/12	Medomak Canning Co.; New Sharon; Corn shop and 2 warehouses lost; 4:30 AM blaze caused $2 Million loss: Assisted Farmington Falls & New Sharon; Only industry in New Sharon with exception of small tennis racquet factory operated by Alan Fitz
	12/25	Peter Mills house; Perham St; David Ferrari & Harold Swain Apartments; House gutted; Total loss
	12/30	Earl Webber house; Maple Ave; Total loss
1961	2/04	John Newcomb; Wilton Rd; House, barn & Ray-Jon Kennels; Total loss
	3/05	Merton Petrie home; Lower Main St; House completely gutted; Total loss
	4/15	Charlie Oliver house; West Farmington; Barn and shed lost; House saved
	8/27	Harry Knox home; West Farmington; Total loss
	12/03	Gladys Bailey farm; Barn and contents lost; House saved; Pumped water from brook at foot of hill; 7 hour fire; 38 lengths of 2 ½ inch hose laid
1962	5/05	Farmington High School; Middle Street; East end of school where equipment room and coaches' rooms located; Heavy damage; 4 hour fire
	5/20	David Fortier; Holley Road; Buildings in flames; Total loss; No water
	7/27	Brown's Restaurant; Main Street; Kitchen destroyed; Dining room saved
	12/07	Hubert Knowlton Roland Dube Apartment: Heavy damage to upstairs and contents
1963	2/22	Mallory Farm; South Strong Rd; House lost; Assisted by Strong, New Vineyard & Farmington Falls
	5/25	Lambert's Filling Station; On the Flat; 2 bays involved from gasoline can spark that ignited
	10/03	Roy Yeaton camp; Allens Mills; Total loss
1964	1/15	Albert Black home; North Main Street; House total loss; 4 year old boy died
	1/16	Colby Ryder home; Farmington Falls; Total loss; Saved home of William Hiscock nearby
	1/17	Elton Powers home; Knowlton Corner Rd; Total loss; Lack of water
	1/18	Dan Chandler home; Lower Main Street; Shed and ell lost; Damage to main house; Gas Explosion; Heavy smoke and water damage
	2/26	Carl Lyon's Garage (Batzel's); West Farmington; Damage to back of garage
	6/20	Barden house; Temple; Total loss
	7/16	House on Starks Road; Assisted New Sharon; Total loss
	8/3	Erland Hamlin; Chesterville; Total loss
	8/3	Tyler Currier home; Fairbanks; Lost barn, saved house
	11/15	Town of Phillips; Several buildings in Upper Village lost
	11/15	John Gravlin home; Strong Road; Total loss

1965	1/01	Bouffard Furniture Store; Main Street; Basement; Heavy damage to building
	1/14	Willard Jackson home; Perham Hill Road; Home total loss
	4/03	Newman Motors; Broadway; Garage & show room destroyed; Explosion blew roof up; Assisted Wilton and Jay
	5/21	Wayne Heath home; North Chesterville Road; Total loss
	8/30	Elden Hall camp; Clearwater, Industry; Heavy damage to interior of camp
	10/25	Brackley's Mill; Strong; Mill destroyed; Pumped water 3 hours to protect nearby properties
	11/24	Charles Glidden house; West Farmington; Barn destroyed; House gutted
1966	2/01	Meisner's home; Webster Road; Kitchen involved; 2nd call same day; more damage
	4/20	Voter Hill Farm; Lower barn totally destroyed; Main buildings saved
	4/24	Stanley Ellsworth house; Knowlton Corner Rd; Cellar and kitchen heavily damaged; Dryer ignited
	8/31	Allen Tracy home; Chesterville Road; Garage lost in fire
1967	4/28	Mission (Gladys Pease; back of corn shop by the river); Small house lost
	5/01	Brick yard; West Farmington; Whole yard in flames; 2 buildings, wood & lumber lost
	9/09	Richard Adams house (old Rackliff farm); Owned by Peter Mills; Back Falls Road; 2-Story; Total loss
1968	1/13	Jack Welch home; Bailey Hill; House destroyed; 2 car garage saved
	2/08	Fairgrounds; New horse barn; Building gutted; Loss $9000; Mechanical equipment lost and 11 race horses succumbed to smoke valued at $1000 each; 10 other horses got out with resuscitator saving one horse; 7:30 PM blaze believed to be caused by heater.
	2/19	Mrs. Geo. Yeaton home; West Farmington; Kitchen in flames; Heavy damage
	3/05	New Vineyard home; Atwood house; Total loss
	3/12	Helen Osborne; Titcomb Hill; 2 barns destroyed
	3/30	Herbert Parlin home; Allens Mills; Barn, ell & house destroyed; High winds
	4/09	Fairgrounds; Horse stable and back of grandstand destroyed; High winds
	5/16	Merritt Averill home; West Farmington; Barn, shed, house; Total loss; 5 hrs.
1969	7/21	Willard Hewey house; House; Total loss
	11/23	Whitney house; Industry; Fire in ell, entering main house; Heavy damage
	12/06	Strong; Main Street destroyed; Assisted Strong, Phillips & New Vineyard

Horses Succumb to Smoke at Stable in Farmington

FARMINGTON — A total of 11 race horses, trotters and pacers, died of smoke inhalation, early Thursday night, when fire broke out in a stable at the Franklin County Agricultural Society Fairgrounds here. Possible injuries suffered by the other 10 horses housed there were unknown late Thursday night.

Firemen from the Farmington and West Farmington departments brought the blaze under control at 7:30 p.m. about an hour after Ernest Danforth, who lives in a trailer about 50 feet from the stable, heard an explosion and saw heavy smoke and flames.

The building, constructed in 1966 and opened for use in January, 1967, cost $18,000. It was expected that the fire caused damage amounting to about half of the original cost. The building was gutted. Each of the dead horses was estimated to be worth about $1,000.

Lost were the following horses: Wilton Wonder, Hard Luck Girl and Jeff Dudley, owned by Murray Smith; Little Joe Ben and Colonel Guy, owned by Town Officer Sheridan Smith; Filly Frisco owned by Albert Bergeron; Magic Switch, Better Wind and an unidentified horse owned by Donald Buchanan; Freddy W. H. and an unidentified four-year-old owned by Llewellyn Bubier.

Of the 10 horses escaping death, one was owned by Buchanan, one by Bubier, three by Danforth, three by Clyde Hathaway, one by Donald Adams and one by Morris Wing. The four trucks and firefighters at the scene were directed by Fire Chiefs J. Bauer Small of Farmington and Richard Kingsbury of West Farmington.

Robert McCleary, a fireman and the president of the Agricultural Society, stated that the fire might have started around a heater. He said the smoke was strong and that a horse could not live long in it. According to reports, a resuscitator was used on one of the fallen horses. The animal stood up and was lead from the stable.

1968 Fire at Fairgrounds
11 Race Horses, Trotters & Pacer died of Smoke Inhalation
Murray Smith, Sheridan Smith Albert Bergeron, Donald Buchanan & Llewellyn Bubier lost horses
Building Loss $9000, Horses Valued $1000 each

Article Courtesy of Don DeRoche and Franklin Journal

1970	4/24	Fred Koch farm; Morrison Hill; Barn & house lost; Returned 3x to scene
	5/01	Merton Edwards garage; Middle Street; Building a total loss
	5/11	Perkin's Place; Industry; House total loss; Assisted Industry
	6/11	Harold Hardy barn; Lightning strike; Barn completely destroyed
	7/02	Conant farm; Lightning strike; Barn completely destroyed
	7/25	Farmer's Union; Basement; Minor damage to building; Some inventory lost
	10/29	Leonard McPhee home; Strong Road; Small barn destroyed
	11/19	Howard Pease house; Back of B&M Corn Shop by bridge; House destroyed; Lester C. Hutchinson died in fire

TWICE A WEEK — TUESDAY AND FRIDAY   AND FARM

VOLUME 119 NO. 39   FARMINGTON.

Main St., Phillips

Norman Field Dies During Raging Fire

Norman H. Field, 56, prominent Phillips businessman, collapsed and died on Main Street in the midst of a raging fire which destroyed all but two buildings of the business section in Phillips on Saturday afternoon.

Mr. Field was born April 25, 1914, in Phillips, the son of Harry H. and Pearl Timberlake Field. He attended Phillips High School and was graduated from Exeter Academy, Exeter, N.H., in 1933 and from Bowdoin College in 1937. He married Jeanne Badger of Rangeley in 1937.

Mr. Field worked as principal statitician for the Maine Employment Security Commission in Augusta until 1951 when he returned to Phillips to assist his father in the management of the Field

(Continued on Page 8)

Rabies Clinic Here Friday

A rabies clinic will be held from 6-8 p.m. Friday, April 16 at the Community Center. The clinic is being sponsored by interested citizens of the area who urge that persons having cats as pets bring them to be immunized. Dogs may also receive the immunization.

Dr. Dana Dingley will administer the vaccine. A minimum charge will be made.

The count of rabied animals is the highest in Maine's history. It is necessary that you protect your pets from this dread disease. Human lives, too, are endangered

FIRE DESTROYS PHILLIPS BUSINESS AREA — This   Saturday afternoon fire causing sev

lin Journal

TON CHRONICLE

TEN CENTS PER COPY — FIVE DOLLARS A YEAR

TUESDAY, APRIL 13, 1971

EIGHT PAGES

Demolished By Fire

Firemen From Six Towns Battle Blaze

Main Street Virtually Wiped Out; Damage Estimated at $900,000

The town of Phillips in North Franklin County suffered a devastating blow on Saturday when fire destroyed all but two buildings in its business district. Firemen from six communities battled the blaze for approximately four and one half hours. This is reportedly the fourth time in the past 100 years that fire has raged in this community of over 1000 population. Loss from the blaze is estimated to be several hundred thousand dollars.

Norman H. Field, owner of one of the business buildings collapsed and died. He and Mrs. Field had been in the process of removing records from his office, before the fire had reached the structure. Due to loss of telephone communications officials were unable to contact the Franklin County Memorial Hospital Ambulance Service. A local ambulance and physicians arrived to aid Mr. Field, but the doctor was unable to revive him.

Fire Departments from Phillips, Strong, Kingfield, Farmington, New Vineyard and Rangeley battled the blaze as pumpers provided water from the nearby Sandy River. Thousands of feet of hose were strung the entire length of Main Street and across the river.

State Investigator Judkins made a preliminary investigation of the fire on Sunday and Monday with County Sheriff Kenneth L. French and other officials made an effort to determine the cause of the blaze, which is believed to have started in the stairway between the second and third floors of the Beal Block.

Firemen saved two buildings in the business district, one housing the town offices and offices by Charles Young, of Dryden.

The Beal Block housed stores, a ladies furnishing store operated by Mrs. Lillian Dilland, a hardware and men's clothing and shoe sales business owned and operated by Herbert Mitchell, of Phillips and the IOOF and Masonic Hall. The fire continued on that side of the business street, and the Field Pulpwood building, which also housed the Blaise Morris on Insurance Agency, now owned by the Currier Insurance Co., of Farmington.

The fire also burned at the other side of the street, gutted the Bates Block, which housed a variety store, Masonic Hall, U.S. Post Office and stores through Edmunds Red and White.

ands dollars. Fire Companies from Strong, New Vineyard, Kingfield.

1971 Fire Main Street Phillips
$900,000 Loss
Photo Courtesy of Franklin Journal

1971	1/04	Barker's Garage; High Street; Paint & storage destroyed; 4 hours; Assisted Wilton
	2/07	Roy Judkin; New Vineyard Road; House destroyed
	4/10	Phillips; Beal Block & Center of Town; Destroyed on both sides of Main St; 10 businesses, 4 family's homes; $900,000 loss; 4½ hrs.; Assisted Phillips, Strong, New Vineyard, Kingfield, & Rangeley: Owner of one of businesses, Norman H. Field, collapsed and died from heart attack.
	7/10	Wilton business block destroyed, Center of Town; Brought under control in 3 hours
	7/28	Richard Barr field and baled hay; Returned to scene 2x on 7/29 to wet hay
	9/11	Search Party for Judy Hand (15 yrs. old); Middle Street to Sandy River & gravel pit; Negative results; Farmington Fair week. Police found body in sawdust pile 12 days later at High Street sawmill which been had searched on 9/11; today UMF's Health & Fitness Center. First murder in Farmington in 40 yrs; Unsolved; one of Maine's cold cases.
	11/09	Federal Row; Industry; Camp in woods destroyed; Chief Forest Allen received medical attention for smoke inhalation and sent home
1972	4/25	Leiby's Store; Farmington Falls; Smoke in basement
1973	2/10	Rachel Chick home; West Mills; House total loss
	4/03	New Sharon; House on Water Street; Total loss; Assisted New Sharon
	8/28	West Mills Church; Struck by lightning; Fire out of control; Total loss
	9/29	Temple Fire Station; Fire caused by down power lines; Heavy damage
	12/29	Caldwell home; Maple Ave; Extensive damage
1974	1/20	Irvin Ditzler home; New Sharon; Extensive damage; Assisted New Sharon
	1/22	Lowell Dubay home; Route 27 New Vineyard; Extensive damage
	1/26	Bernard Allen trailer home; Chesterville; Destroyed; Allen's son died
	3/15	Sampson Apartments; Main St; Major damage to shed, ell gable & attic
	8/4	Old Jennings farm; Mosher Hill Road; Old barn destroyed
	9/01	Jim Martin Subaru Garage; Wilton Rd; Shop heavily damaged
	9/28	Munson farm; Wilton; House and ell involved; Assisted Wilton
	11/24	Dr. Swallow farm; South Strong; Heavy damage to house; 3 hours
1975	1/16	Vernon Hiscock home; Farmington Falls; Heavy damage to house; 2 hours
	1/18	Ronald Smith house; Russells Mills Road; House lost; Temp -18; Water hauled; Trucks, nozzles and hoses froze; Assisted by Temple
	2/03	Lindy Foss house; Mosher Hill; Temp - 15; No Water; Hoses froze; Lost house
	4/02	Drummond Hall Block destroyed; 3 businesses lost; $100,000 loss; Assisted by Wilton, Temple called; 10 hours; Returned to scene 3x
	9/06	Steve Palichak (NJ) camp; Industry; Destroyed; Assisted by Industry
	10/12	Vacant store and Robert Stanley house; Chesterville; Destroyed; Assisted by Chesterville and Farmington Falls
1976	4/15	79 North Main Street; Jean Hunter, mother, and Troy Hunter, 2 year old son died in fire; Main house destroyed

March 2, 1975 — Fire on Broadway

1975 Fire Drummond Hall Block on Broadway
3 Businesses,
$100,000 Loss
Fireman Bob McCleery Lower Left

1977	4/22	Tempesa camp; Clearwater; Industry; Destroyed
	7/24	Larry Wattles; New Vineyard; House heavy damage; A. Wattles overcome by smoke rescued from 2nd Floor
	8/17	Richard Smart house; Industry; Near Shorey Chapel; Heavy damage
	8/23	Macomber Mill (Lugar & DiStefano); West Farmington; 2nd Fl. block storage
1978	1/05	Neal Yeaton; Lower Main Street; Timbers under fireplace in basement
	2/25	Accident at Rotary near bridge responding to call; Clyde Ross, Jack Bell & Terry Warren taken to Franklin Memorial Hospital with injuries
	2/28	Ron Jahoda house; Farmington Falls Rd; 1st Floor; Heavy damage; $20,000
	3/5	Wallace French (Robert Grover); South Chesterville; Total loss
	4/12	Olive Gilbert Taylor; Whittier Rd; Kitchen and dining room; Heavy damage
	11/23	Everett Vining & D. Lewiski; West Farmington; Starling Street; Total loss
	12/05	Kenneth & Sherry Breton house; Lucy Knowles Road; Total loss
1979	1/24	Vernon Peacock; Route 156; Buildings; Total loss
	2/01	Woodrow Adams; Route 43; Mobile home; Heavy damage
	4/13	Paul Jackson house (Gary Black family tenant); 75 N. Main St; $20,000 loss
	5/11	Roy Hazzard (Old Geo. Thomas farm); Chesterville; Stand by at Station; Engine #3 stand by at fire; Total loss
	10/15	Mike Graham apartments (Old VIS Hall); Bridge St; West Farmington; Heavy damage
	12/01	Smith Mill; New Vineyard; Total loss
	12/18	Doug Swain house; East Wilton; John Pollard died; Assisted Wilton

Structure Damaged Heavily By Fire

FARMINGTON — Facts from Fire Chief Robert L. McCleery Thursday verified the first report of late Wednesday evening that the story and a half wooden frame house that was extensively damaged by fire on the Back Falls Road was unoccupied. However, contrary to the first report received during the height of the blaze, the vacancy had been only since February.

McCleery said that Mr. and Mrs. Paul Gilbert and family had resided in the home until February when the family moved to Livermore Falls. The house is part of the property of Gilbert's late father, Frank Gilbert. A daughter, Mrs. Olive Taylor with her husband and children, reside next door in the home occupied by her father, Mr. Gilbert, until his death.

Mrs. Taylor apparently noticed a light in the vacant house and notified police about 10:30 p.m. that she could see a flickering light in the second floor window. The house is located nearly a quarter of a mile from U.S. Rte. 2 and State Rte. 4 from where hoses were laid from a fire hydrant near the main traveled highway.

McCleery stated that at least 20 firemen responded to the call, bringing the Squad Truck and four engines. Hawthorne Ambulance was also at the scene and the Central Maine Power Company dispatched a crew and truck to the area.

McCleery said it was discovered that the fire started at the foot of the stairs in an electrical box. The fire was drawn up the stair well and demolished the second floor, burning through the roof which was sagging by the time the fire was extinguished. He said very little fire damage was noted on the first floor of the home.

The firemen were just returning to the fire station from a training session with Dick Cadwell from the State Department Fire Training Service when the house was called in. The men had been working on a simulated major disaster reported on Front Street in the vicinity of the Farmington Farmers' Union Store. The exercise in which Wilton and Temple firemen also participated, was planned so that the firemen were hampered by various problems in covering the disaster to the best advantage. After each phase of the simulation the firemen would congregate to compare notes and discuss what they had done wrong. Firemen from Industry observed the training session.

McCleery said Thursday he was "well pleased" with the training session and also with the work of the firemen at the real fire which followed.

Farmington Police and State Police directed traffic on Rtes. 2 and 4 and kept the Back Falls Road clear of vehicles for the fire equipment. The hoses were strung across the busy highway holding up moving vehicles for a time.

The firemen had just about gotten back to the fire station when the car fire was called in.

1978 Gilbert
Vacant House Fire
Back Falls Rd.

Article and Photo Courtesy Sun Journal

1978 Vining
Warehouse Fire
Thanksgiving Day
Total Loss

Photo Courtesy Town Farmington

Robert L. McCleery III

Fire Guts Old Wilton Academy

By RUTH ADAMO

WILTON — Fire completely destroyed a portion of Wilton Academy Saturday afternoon, and did considerable smoke and fire damage to the rest of the old building. The rambling building, which was used as a junior high school for the sixth, seventh and eighth graders of Wilton, consisted of a main portion and two annexes built in later years.

Flames were observed at the main entrance of the academy at about 2 p.m. Saturday by eyewitnesses who ran to a house to call in the fire.

Thousands lined the streets and watched as the fire, which reportedly began in the area of the main entrance, at the back of the school, swept through the original part of the Academy. Many of those watching had been graduates of the school when it was a four-year high school (from its origin in 1867 until 1962) and emotional reactions to the sight were common.

The Academy had originated as The Old Meeting House for the use of three churches in about 1830, and was converted into a school in 1866-67.

Fire departments from Jay, Farmington, East Wilton and East Dixfield were called, and backup units from Temple and Farmington Falls assisted at the Farmington station.

A total of 114 firemen had battled the blaze, pumping over 400,000 gallons of water, before the worst of the flames were extinguished late Saturday evening.

Seven firemen were treated for smoke inhalation or other injuries, and three were hospitalized. Cary Pond, a Hot Shot (junior fireman) was treated at the scene and taken to the hospital, then released. Also treated at the scene was another Hot Shot, Michael Donald. This group of young firemen fought a "big fire" on Birch Street, according to

Continued on Page 14 Column 3

EAGLE ABOUT TO FALL — Wilton Academy's belfry topped by the eagle, symbol of Academy high school teams, crashed in flames Saturday afternoon seconds after this picture was taken. More fire news on Page 15. (Yeaton Photo)

1980 Fire Destroys Wilton Academy

Top: Article Courtesy of Lewiston Sun Journal

Bottom: Article Courtesy Waterville Morning Sentinel

Flames Destroy Wilton Academy

By JEAN KING
Sentinel Correspondent

WILTON — Wilton Academy, a 151-year-old local landmark where generations of local residents went to school was destroyed in a major fire Saturday.

Seven firemen from five area departments were treated for minor injuries and smoke inhalation at the blaze, first spotted at 2:14 p.m. by a newspaper correspondent.

The main building, which once was used as a church, was leveled in the fire. An addition where classrooms and a science laboratory were located was damaged but still standing Sunday.

THE WIND-SWEPT blaze quickly engulfed the structure at the junction of Depot Street and Old Route 2, despite the efforts of firemen from Wilton, East Wilton, East Dixfield, Jay and Farmington.

Wilton Fire Chief Wade Atwood, who took charge of the fire fight, said Sunday the cause of the fire is still under investigation.

It was unknown how long the fire had

Wilton Landmark Mourned: Page 13

been burning, he said, when it was first spotted.

Mrs. Jean King of Wilton, who turned in the alarm and who lives near the academy, said two girls came to her door at about 2 p.m. Saturday to report there was smoke coming from the three-story building.

Mrs. King said she went to the academy, saw smoke coming from around the main entrance and called firefighters from a nearby home.

THE FIRST UNITS, from Wilton, were on the scene in minutes, but by then flames were already visible through the windows.

As the fire grew in intensity, Mrs. King said, windows began breaking out and flames appeared everywhere.

Ladder and tank trucks were pressed into service as more than 100 firefighters, the majority of them volunteers, battled the blaze.

Among those treated at Franklin Memorial Hospital in Farmington were Cary Pond and Gregg Oakes, both of Wilton, who suffered smoke inhalation. Oakes was admitted overnight and discharged Sunday.

FIVE OTHERS from the area were also treated at the scene for smoke inhalation and cuts and abrasions. There were no serious injuries.

Hundreds watched the spectacle, some of them weeping, as the fire raged out of control.

The academy, once a private school, had been last used as a middle school by School Administrative District 9. SAD 9 officials met later Sunday to determine where to send the displaced youngsters.

Firemen were still at the remains of the church-turned-school Sunday as lines of cars filled with onlookers moved past.

Chief Atwood estimated 500,000 gallons of water were used in the fire fight. He commended cooperating fire departments and emergency personnel who participated in battling the blaze.

The cupola atop Wilton Academy's main building totters into an inferno of flame despite efforts by more than 100 firemen to save the 151-year-old landmark. The fire razed the one-time church and damaged an addition containing classrooms.

Sentinel Photo by Jean King

1980	2/24	Greg Wilson; West Farmington; 2nd Fl.; Damage to apartment

1980 2/24 Greg Wilson; West Farmington; 2nd Fl.; Damage to apartment

3/12 Harold Lothrop; North Chesterville; House and contents; Total loss $45,000

5/10 Wilton Academy; Main Street, Wilton; Main School Building leveled; Annex saved, but heavily damaged by water & smoke; Over $1 Million loss; Farmington, East Wilton, East Dixfield, & Jay assisted Wilton; Temple & Farmington Falls on stand-by. 114 Firemen; 31 from Farmington on scene; Firemen remained on scene from 2 PM Sat. till 9 AM Mon; 400,000 gals. water pumped; 7 firemen treated smoke inhalation; 3 hospitalized; one (1) Farmington fireman treated with oxygen on scene

5/25 Elmer Eaton; South Strong Road 8/29 Robert Stevens house & apartment (L. Richardson & Nancy Hall); Lake Ave; 2nd Fl. Heavy Damage; $45,000 loss

9/8 Sturges Butler; In Town; few acres, storage barn and damage to building and stored equipment

1981 1/02 Leo Karkos; Chesterville Center; House total loss $50,000

1/04 Robert Hjort; Industry Road; New Sharon; $45,000 loss

1/05 H. Bruce Judgkins; New Sharon; Building under construction

1/23 Mim Simcock house; Holley Farms; Holley Farms Rd; Total loss

2/02 Rescue 2 men from Sandy River; Canoe capsized; Flood, ice jam

3/08 Jackie La Chance; Mills Park; Mobile home; Total loss

5/08 Merle Norton; New Vineyard Basin; Barn total loss; Assisted New Vineyard

6/12 Fairgrounds; Cattle shed; Minor damage

7/25 Janice Magno; Lower Main Street; Barn and ell; $10,000 loss

1982 1/17 Earl Luce; Upper end Holley Rd; 0 degrees, high winds at 2:25 AM; 7 hrs.; House total loss $50,000

4/05 Joan Colleen Hickey; 70 High St & Lake Ave; 4:20 AM fire; Living room & bedroom; Joan C. Hickey (Age 45) & daughter Amy Bicknell (Age 16) died in fire from smoke inhalation; L. Golbel Patten, Ms. Hinckley's father and step-mother escaped by going out bedroom window; Loss of Life; Buildings $75,000 loss

9/20 Albert Tyler (Nancy Tyler & Larry Norton occupied): RT 4, Strong Rd; Living room and bedroom badly damaged; Elisha Mae Tyler (Age 2) died from smoke inhalation; 7 hours; Loss of Life; Buildings $40,000 loss

11/7 Brackley's Service Station; Strong; Heavy damage; Assisted Strong

11/18 Donald Clark; West Farmington; 2nd Fl. Heavy damage

12/09 Kenneth Hoar; Town Farm Rd; Assisted Temple & Wilton; Total loss

1983	1/02	Clyde Pingree; Stanley Rd; New Vineyard; Kitchen and ell; Heavy damage; 6 hours; Assisted New Vineyard
	6/24	Richard Davis barn; New Sharon; Wet hay, hay loft smoking; Assisted New Sharon
	8/17	United Timber (Starbird); Farmington Falls Rd; Office, showroom, and storage buildings; 15 hrs.; Assisted by Wilton and Temple; $400,000 loss
	10/07	Woodrow Adams; Industry Rd; Garage & storage shed; Total loss; $5000
	10/15	Tri-County Mental Health; 1st Floor Hall; 2nd Floor Office
	10/27	Althea Nida: Water Street; W. Farmington; 2 story apartment house; 2nd Floor damaged; $50,000 loss
	11/05	Dana Dingley summer house; Industry; Total loss; Assisted Industry
	12/17	Country Haven Nursing Home; Temple; $50,000 loss; Assisted Temple

1983 United Timber (Starbird Lumber)
Fire Destroys Office, Showroom and Storage Buildings
$400,000 Loss

Photo Courtesy Don DeRoche

1984	2/09	Percy Harris home; New Sharon; Total loss; Assisted New Sharon
	2/13	Apartment house; School Street, Wilton; Assisted Wilton
	2/28	Robert Fenn barn; North Chesterville; Barn & 28 sheep lost; Assisted by Temple & Chesterville
	3/27	Fred Wetmore trailer; Lambert's Trailer Park; Fairbanks; Heavy damage
	5/02	Leroy & Mildred Hammond; Hammond Rd. off RT 133, Both died; Call came in at 4:48 AM; Loss of Life; Buildings lost; Assisted Wilton
	5/02	Ted Marcou house; Town Farm Rd; Vacant house; Arson

| 10/07 | Truman Taylor house; Dutch Gap Rd; Chesterville; Heavy damage to 2nd Fl; Assisted Chesterville |
| 12/02 | Dr Helmet Bitterauf barn; Russells Mills Rd; Total loss; Assisted by Temple & Chesterville |

1985

1/09	Albert Davis trailer; Sunrise Trailer Park; Total loss
2/10	Golden Dragon Restaurant; Mt. Blue Shopping Center; Heavy damage; Assisted by Temple
3/05	College Apartments; 44 High Street; Basement & laundry; Minor damage Arson
3/21	P. Dorman Store & Apartment; Main St, Wilton; $50,000 loss
3/22	Dr. Robert Martin; Cape Cod Hill, New Sharon; $40,000 loss
4/23	Bruce Nile home; Temple; Total loss; Assisted Temple
7/22	Dr. Bardo Office; Depot St, Dryden; $10,000 loss; Assisted Wilton
9/14	Bev Beisaw house; Quebec St; Fire in laundry area; $48,000 loss; Assisted by Wilton & Temple
12/08	Mike Sayward & Pete Richardson; 2 ½ story house; New Vineyard; Total Loss; Assisted New Vineyard
12/13	Arthur Whitney house (Dennis &Agatha Mills); Marvel Street, West Farmington; $20,000 loss

1986 Tracy Fire Seamon Rd

1986

1/02	Tom Williams pig barn & hay storage; RT 27; Several pigs & barn lost; 18 hrs.; $40,000 loss; Assisted by New Vineyard, Temple, New Sharon & Wilton
2/14	Larry Tracy house (Peter Mills owned); Seamon Rd; Total loss; $15,000
3/09	Sweetser's Mill, RT 43 Temple Rd; $45,000 loss; Assisted by Temple & New Sharon

3/12	Jon & Barbara Luce; Lower Main St; 2 ½ story house, 2nd floor & attic damaged; $20,000 loss
3/22	Constance Kolreg mobile home; Back Temple Rd; $20,000 loss; Assisted Temple with tanker shuttle
5/03	Robert Hunter house; Farmington Falls by green bridge; 2½ story house lost; $50,000 loss; Assisted by Temple, Chesterville & Wilton
8/05	John & Kathy Dorr; Tater Mtn. Rd; Temple; Assisted Temple
9/14	Richard Davis barn & silos; New Sharon; 180 ft. barn and 2 silos lost; $60,000 loss; Assisted New Sharon, Temple, Chesterville, Industry, Wilton & Strong

1987

1/02	Robert Searles house; Barker Road, New Vineyard
1/17	Frances Orcutt; Temple Rd; Garage & storage building; $40,000 loss
2/25	R. Searles house (old H. Seamon house) Knowlton Corner Road; Total loss
3/17	2 Story Apartment House & Stable; Water St, West Farmington; $30,000 loss
4/02	Lena Ladd house; Farmington Falls; Flood caused malfunction of gas heater; Total loss
6/22	Apartment House; Weld St, Wilton; Assisted Wilton
9/06	Arthur Porter Mercantile Building; RT 2, New Sharon
12/22	Daniel Dubay garage; South Strong Rd; $25,000 loss
12/31	Jr. Dubay barn; Barker Rd, New Vineyard; Barn destroyed; 4 animals lost
12/31	Leland Searles home; Main St, New Vineyard; Total loss; Assisted New Vineyard

1988

1/25	Smith Road house; New Sharon; Total loss; Assisted New Sharon
2/05	Ron Petrie house; Gridiron Hill, Industry; Assisted Industry
2/08	Marilyn Shea Apartments; High St; 5 APTS; $35,000 loss; Assisted by New Sharon
3/12	Tom Williams Apartments; 120-121 Main St; 8 APTS; $75,000 loss; Arson; Assisted by Wilton & New Sharon
3/23	John Brown Jr. house; Walker Hill, Wilton; Assisted Wilton
3/25	Tom Gill house; North Chesterville; 2nd Fl.
6/25	Keith Howard house; RT 27; 1 ½ story house; Lightning strike; Total loss; Assisted by Temple, New Sharon, New Vineyard & Strong
9/23	Margaret Nottage; Starling St, West Farmington; Cooking fat; $5000 loss

1989

1/02	Richard Doan mobile home; Intervale Rd, behind diner; Total loss
1/23	Jessica Jahnke & Maria Stickle home; RT 27; Moderate damage
2/28	St. Pierre home; Orchard Drive, Wilton; Assisted Wilton
5/21	James Emerson home; RT 27 N, New Vineyard; Assisted by New Vineyard, Industry, Temple & Chesterville
5/27	Harold Kyes home; Industry; $60,000 loss Assisted Industry
6/28	Northeast Wood Turning Mill; West Farmington; Hot shavings; $40,000 loss; Assisted by Temple
7/22	Northeast Wood Turning Mill; W. Farmington; Sawdust bin; Minor Damage
9/22	Robert Davis; Town Farm Rd; Assisted by Temple, Wilton & New Sharon
12/12	Mike's Auto Body & Apartments of Jordan, Roux, Egars; 40-42 Broadway;

Heavy damage; 6 hrs.; $50,000 loss; Assisted by Industry, Temple, Wilton & New Sharon

12/24 Emma Spearin home; RT 43; Total loss; Assisted by Industry, Temple & Chesterville

12/30 Diane Beisaw home; Church St, New Vineyard; Assisted New Vineyard

1990

1/24 UMF Lockwood Hall; Lounge & basement; Suspicious fire; $10,000 loss

2/02 New Attitude Clothing Store (Lornie Carter, prop/ Dennis Pike, bldg owner); Wilton Rd; $125,000 loss

2/05 Gordon McBean property; Wilton; Assisted Wilton

2/10 Timberland Inc Garage; Strong; Assisted Strong

3/02 Kingfield Wood Products; Assisted Kingfield

3/06 Susan Morse home; Davenport Hill; Jay; 2 Fatalities; Assisted Jay

3/17 Basil Durrell home; 2 Sunset Ave; $10,000 loss; Assisted by Wilton

4/16 Christopher Muise apartment (M. DiDonato owner); 14 Court St; 4 Apt. lost; $10,000 loss

4/26 Timberland Wood Products Mill; North Main St, Strong; Assisted Strong

5/12 Maiden Lane Apartments (Tom Williams owner); 3rd Fl. Apt; Pan on stove;$30,000 loss; Assisted by Wilton, New Sharon, Temple, Industry & Strong

5/26 Cliff Norton house; Porter Hill Rd; Basement

6/02 Northeast Wood Turning Mill; West Farmington; Spark in wood turnings; $3000 loss

6/05 Bruce Whitney barn; Lucy Knowles Rd; Barn lost; $8000 loss

8/21 RST truck roll-over; Foot of Hill St & Intervale Rd; Haz-Mat; 15 hrs.; Assisted by Mutual Aid FD, International Paper, DEP & RST HazMat Team

10/19 Mike Bolduc house; Cushman Dr.; $30,000 loss; Assisted by Strong

11/30 Edward Sheetz barn; Crowell's Pond Rd; Barn lost; Assisted by Chesterville

12/30 Fatal Accident of Angela Butterfield at Tom Nelson farm; Webster Rd, New Sharon; Caught in tractor power take-off; Assisted by Industry & Wilton

1991

3/22 David & Teresa Currier home; RT 4 North; Suspicious; $40,000 loss; State Fire Marshal called

4/14 Weld Rd brush fire; Wilton; 4 hours; Assisted Wilton

4/15 CMP power line fire; Davis Rd; 2 acres burned; 3 hours; Assisted by Wilton & Industry

4/22 Ralph Searles Jr. home; RT 2 & 4 West Farmington; $40,000 loss; Assisted by Wilton, Industry & New Sharon

5/01 Glen Stowe farm; Outdoor power lightning strike; $15,000 loss

6/13 Forster Manufacturing; Depot St, Wilton; Sawdust bin; Assisted Wilton

6/14 Fred O. Smith Mill; New Vineyard; 4 hours; Assisted New Vineyard

6/15 Phil Sweetser mobile home; Temple Rd; $15,000 loss; Assisted by Temple, Chesterville & Industry

6/15 Mrs. Ralph Dunton; Main St, New Sharon; 4 hours; Assisted New Sharon

8/07 Fatal Accident of Herbert Hoffler; Wilton; Truck with propane tanks

12/10 Forest Greenman storage shed with ATV; Knowlton Corner Road; $3000 loss

1992	1/19	Doug Becker Apartment House; 11 Quebec St; 4 Apts. occupied; 4 hours; $40,000 loss; Assisted by Industry & Wilton
	1/21	Charles Lugar home (R. Kenny tenant); Wilton Rd; $5000 loss
	2/12	Paul Friend house; Industry; 6 hours; $60,000 loss; Assisted Industry
	3/24	Warren Elliott mobile home; Ridge Rd, Chesterville; Assisted Chesterville
	5/03	Norman Bean home; Industry; $3000 loss; Assisted Industry
	5/12	James Hunter mobile home; Sandy River Park; $22,000 loss; Assisted by all Mutual Aid Fire departments.
	6/18	Thayden Farrington farm; Wilton; Storage Building; Assisted Wilton
	6/21	Raymond Weirs home; Wilton; Lightning strike; Total loss; Assisted Wilton
	8/21	John Distefano home; Porter Hill; 5 hours; $30,000 loss; Assisted by Strong, Temple, Industry, Wilton, Chesterville, New Vineyard & New Sharon
	9/05	Knowlton-McLeary Printing; $25,000 loss; Assisted by Temple, Wilton, Industry & New Sharon - using Water-Vac
	9/09	Treatment Plant; Box trailer storage; $25,000 loss
	10/10	Steve's Market; Dryden; 7 hours; Total loss; Assisted Wilton
	10/14	Sanborn Hill house; Chesterville; 5 hours; Total loss; Assisted Chesterville
	11/21	James Bealeau house; RR1 East (New Sharon line); Total loss; Assisted Industry
1993	1/06	O'Brian's Radio Shack; East Wilton; Assisted Wilton
	3/17	Clyde Nile house; Morrison Hill Rd, West Farmington; $25,000 loss
	3/22	Country Manor Restaurant (Ronald Rackliff) Farmington Falls Rd; Assisted by Chesterville, New Sharon & Temple
	4/18	Arthur Lambert house; Morrison Hill Rd, West Farmington; $80,000 loss; Assisted by Chesterville, Industry & Temple
	4/29	Tom Williams storage shed; Front Street; Vacant; $10,000 loss
	5/16	Bonnie Annadale Apts.; 121 Main St; $15,000 loss; Assisted by Wilton; Fireman Mike Bell broke arm
	5/17	Ferrari Bros. Clothing Store (Marty Pike owner); Main Street & Broadway$350,000 loss; Assisted by Chesterville, Industry, Jay, Wilton, Temple & New Sharon
	7/27	Ronald Saultes mobile home (Steve Mahar); Holley Rd; $4000 loss
	8/02	Pam Whitney house (George Lowell); RT 27, New Vineyard Rd; Total loss; Assisted by New Vineyard
	8/05	Ruth Osgood Apartment House; Fernald Street, Wilton; Assisted Wilton
	9/24	Wes Moody barn; RT 27, New Vineyard; Minor loss
	10/26	Inn Town Lounge: Main Street, Jay: 5 hours; Assisted Jay & Wilton
	11/12	Northeast Woodturning; West Farmington; wood turnings; $4000 loss
	12/05	Gary Fletcher house; Lake Rd, New Vineyard; 4 hours; Total loss
1994	1/01	Northeast Woodturnings; West Farmington; Wood turnings; $3500 loss
	1/02	IOOF Building; State Theater; Water freeze, sprinkler system; $50,000 loss
	1/15	William Marcous house; McCrillis Corner, Wilton; Total loss; Assisted Wilton
	1/20	Beef Barn (Clinton Brooks); RT 133; Total loss; Assisted Wilton & Jay

	1/26	Fred Quirion house; Depot St, Dryden; Total loss; Assisted Wilton
	4/26	UMF Apartment House; 10 School St; No power, using candles; Lost tapestry; $4000 loss
	5/10	Water Street barn; West Farmington; Fire under barn from 9 year old playing with lighter; Minor damage
	5/12	Mt. Blue Technical Center; Carpentry shop bathroom; Minor damage
	5/22	Steve Ratley Apartment House; 79 No. Main St; 5 apts. lost; $30,000 loss; Assisted by Wilton, Temple, Industry & Chesterville
	7/19	Leo Stevens house; Industry; Assisted Industry
	7/31	Kelley Dexter; Freeman Garage; Strong; Assisted Strong
	9/17	Farmington Motel (Scott & Dan Adams); Farmington Falls Rd; Minor Damage
	9/26	Thayden Farrington; Wilton; Barn & storage; Total loss; Assisted Wilton
	9/30	D & D Ridley home; Mason Rd, New Sharon; Total loss; $100,000 loss
	10/28	Connie Brann home; Lucy Knowles Rd; $10,000 loss
	10/30	Hosmer Edwards, American Legion; Wilton; Heavy damage; Assisted Wilton
	12/03	David & Linda Harris home; Old RT 2, New Sharon; Heavy damage; Assisted New Sharon
	12/12	South Carthage; Wilton; Structure fire; Assisted Wilton
1995	1/09	Cecil Wheeler Sr. home; Borough Rd, Chesterville; Total loss; $50,000 loss
	1/14	Robert Margise; Ridge Rd, Chesterville; Rabbit barn; $10,000 loss
	2/02	Jon Davis home; Granite Heights; Ashes in cardboard box; $20,000 loss
	3/10	Terry Moore home; Morrison Hill Rd, West Farmington; Minor damage
	4/09	Maine Wood Products (Earl Fletcher Mill); New Vineyard; $50,000 loss
	4/10	Bruce Frost home; RT 156 North, Wilton; Assisted Wilton
	5/09	Northeast Wood Turning; West Farmington; Boiler, sawdust bin; $10,000 loss
	5/17	Shadage Road, New Sharon; Total loss; Assisted Industry & New Sharon
	6/20	J. J. Newberry (Tom Mellen); Heavy smoke from furnace; $36,000 loss
	7/01	Sandy River Farm (L.H. York); Lightning strike; 7 hours; Barn and most of house lost; $1,250,000 loss; Assisted by Wilton, Chesterville, Industry, New Sharon, Temple, Jay & Livermore Falls
	8/21	Gerard Williams Law Offices & Apts.; Broadway; $6000 loss
	9/03	Northeast Wood Turning; West Farmington; Sawmill & sawdust bin; 3 hrs.; $10,000 loss; Assisted by Industry, Temple, New Sharon, & Wilton
	10/25	William Shadlow Jr. home; RT 2 East; Interior damage; $20,000 loss
	11/12	Wilbur house; Borough Rd, Chesterville; Total loss; Assisted Chesterville
	11/25	Ted Fisher; Hill Rd, Chesterville; Storage shed lost; $2500 loss
	12/05	Rebecca Lloyd home; RT 27, New Vineyard; $50,000 loss; Assisted New Vineyard
	12/05	Jill Swett home; RT 43, Industry Rd; $10,000 loss; Assisted Industry
	12/16	Orchard Drive, Wilton; 2 hours; Assisted Wilton
	12/22	Natalie Fitch; Weld Rd, Wilton; Home & garage damaged; Assisted Wilton

Top: 1995 Sandy River Farm- L. H. York Fire
Back Falls Rd
Lightning Strike
Barn & Most of House, $1,250,000 Loss

Bottom: 1996 Jesse Crandall Apartment House Fire
Marvel St., West Farmington $30,000 Loss

Photo & Article Courtesy Don DeRoche & Franklin Journal

Fire Damage

Extensive Damage - A 150-year-old house on Marvel Street, West Farmington, was extensively damaged in a fire Sunday night. Firefighters responded from Farmington, Temple, Industry and Wilton, and attributed the blaze to a plumber's torch that had been used in a first floor bathroom earlier in the day. The home, owned by Jesse Crandall Sr. of Mercer, was occupied by his son, Jesse, and roommate Matthew Paradis, both University of Maine at Farmington students. All of their belongings and those of another tenant who had been in the process of moving in were lost in the fire. Estimated damage to the home is about $30,000. Some of the loss was covered by insurance. (Photo by Greg Davis)

FJ Sept 06, 1996 DED

1996	1/19	Devil's Elbow Dance Hall (Carroll Howard); Strong Rd; Total loss
	2/03	Kevin Vining garage (Dan Vallaci, operator); Hill St; Assisted by Industry & Wilton
	2/12	Dennis Fetterhoff house; Zion's Hill, Chesterville
	2/14	Roger Smith house; New Sharon; Total loss; Assisted New Sharon
	4/09	Mike Kemp house; Industry; Assisted Industry
	4/15	Tony Moore house; RT 156, Chesterville; Assisted Chesterville
	5/24	Gloria Tuttle house; Main St, New Sharon; Total loss
	7/16	Thayden Farrington barn; RT 156, Jay; Lightning strike; Assisted Wilton
	8/06	Mark & Tawna Righter; RR 2, Chesterville; Total loss; Assisted Chesterville
	8/12	Animal Shelter; Industry Rd; Basement; $15,000 loss; Assisted Industry, Wilton & New Sharon
	9/02	Jesse Crandall Apt. House; Marvel Street, West Farmington; $30,000 loss; Assisted by Industry, Wilton & Temple
	9/06	Knowlton-McLeary Printing (Peggy Hodgkins); Church Street; $50,000 loss. Assisted by Wilton & Industry
	11/16	Hippach Field Grandstand; Intervale Rd; Arson by 2 juveniles; $20,000 loss
	11/17	James Smith's field; Morrison Hill Rd; West Farmington; Tall grass; No permit
	11/26	New Sharon Motel (Garry & Ellen Mayo); RT 2 & 27; Total loss
	12/22	Daniel Winters home; Cushman Dr, RR3, off South Strong Rd; $35,000 loss
1997	2/01	Elizabeth Chandler house; 5 Belcher Road; $25,000 loss
	2/06	Mt. Blue High School; Seamon Rd; Waste basket - Student; $400 loss
	2/07	Snowmobile Accident; Perham Hill Rd; Robert Adams & Andrew Bailey injured
	2/24	Richard Teele (Eugene Harris, owner); 72 Middle St; $40,000 loss; Assisted by Wilton & Industry
	3/03	Mills Law Office (Peter Mills, owner); Roof of Newberry Block; Minor damage; $200 loss; Suspicious
	3/05	Accident; Back Falls Road; 2 people, male & female, injured
	3/21	Accident; Shop&Save; Truck ran a red light; Serious injury; Wm. McKinley
	4/09	Brian Rebert; Sunset View Estates; 4 qts. oil on stove to warm; $20,000 loss
	4/10	Donald Smiley (Edith Smiley); Knowlton Corner Road; Chimney; $10,000 loss
	4/16	Rescue; Assist New Sharon in river rescue; 3 men in canoe capsized
	4/29	Janet Mills house; Fernwall St, Wilton; Heavy damage; Assisted Wilton
	4/30	Accident: Cushman Drive, South Strong Rd; Amber Collins; Serious personal injury; Ran in front of pick-up truck
	5/02	Accident: RT 2, East Dixfield; Carol Soucie; Fatally injured; Car and tractor trailer truck head-on collision

6/14	Connie Hunter mobile home; RT 156; $15,000 loss; Assisted by New Sharon, Chesterville & Industry
6/22	Lance McNally home; Back Falls Rd; Lightning strike; $25,000 loss
7/02	Jay Williamson house; Farmington Falls; Used propane torch to burn grass; House damaged; $2000 loss
7/29	Rescue; Removed 14 year old girl from top of Tumbledown Mountain
8/10	Eugene Watt vacant house; 15 Sunset Ave; Fire Marshal called; $35,000 loss; Assisted by Industry, Temple, New Sharon, & Wilton
8/19	Accident: RT 2 & 27, Farmington Falls; Jon Manley & Ruth Bond (VT); Fatally injured; Car head-on into R. Carrier Tractor-Trailer
10/02	Accident by Town Garage; Chesterville Hill, Farmington Falls; Leslie Tripp; Fatality; Car struck another automobile
10/03	Accident; Staples Rd, New Vineyard; Car rolled over into brook; Fatality; Young lady drowned
10/13	Mark Gray tractor; RT 43 East; Attempted Arson; $2000 loss
11/18	Kim & Rebecca Atwood house; New Vineyard; Total loss; Assisted New Vineyard
12/25	Accident: High & Broadway; 2 vehicles; J. Benjamin & J. Campbell serious injuries
12/28	Neree Simoneau home; Lake Road, Wilton; Heavy damage; Assisted Wilton

1998	1/01	ICE STORM; Trees on power lines;
	1/08	10 calls over 8 days;
	1/08	Accident; Roll-over on RT 27
	1/09	Hauled water to Charles Hubbard farm for cattle
	1/10	Trees on power lines, 7 calls; Hauled water to Charles Hubbard farm
	1/11	Hauled water to Charles Hubbard farm
	1/12	Accident: RT 4 & 27; Tractor-Trailer roll-over; Load of lumber
	1/19	Wing Farm; Industry; Mutual Aid; Minor damage
	1/28	Accident by Jack's Trading Post; 2 vehicles involved; Minor damage
	1/30	Accident RT 2 & 4 by Fabric Inn; Glenn Stowe received minor injury
	1/30	House fire; Perham Street; Smoke
	1/31	53 Total Calls for Month of January
	2/03	Randy Gibbons 2 story house; Maxwell Rd (Varnum Pond Rd), Wilton; Total loss; Assisted Wilton & Temple
	2/10	Josh Sinclair house trailer; New Vineyard; Total loss; Assisted New Vineyard
	2/12	Peter Barton house; Maple Ave; Chimney fire
	2/13	Accident; Extrication of Cameron Bopp; Corner RT 4 & South Strong Rd; Vehicle struck by log truck driven by Vernon Hall
	2/28	Responded to 10 accidents, 3 vehicle fires, several minor calls; Total 29 Calls for Month of February
	3/01	Andrew Bailey house; Bailey Hill Rd, New Sharon; Structure & Chimney; Assisted New Sharon
	3/30	James Bouffard warehouse; 72 No. Main St; Warehouse & storage structure; Lightning strike; $100,000 loss; Assisted by Industry, Temple & Wilton

3/31	Responded to 3 accidents, 2 vehicles, several misc. calls; Total 20 Calls for Month of March
4/15	Maurice Comeau; Stanwood Park Cir.; Grass fire; Hot coals from brush pile
4/29	Transfer Station; Munson Rd, Wilton; Grass & woods fire; Assisted Wilton
4/30	Responded to 3 accidents, 2 debris fires, 2 minor house fires; several misc.; Total 18 Calls for Month of March
5/12	Accident by Brackley Farm; Strong Road, Extrication; Double Fatalities; Harold & Gladys French
5/29	Lawrence Yeaton; Back Falls Rd; Lightning strike; Hay wagon on interval; Steve Yeaton taken to Franklin Memorial Hospital; Hay wagon total loss
5/31	Responded to 4 accidents, 1 Mutual Aid to New Sharon; several misc. calls; Total 16 Calls for Month of May
6/18	Farmington Chip Plant (Currier Trucking); Town Farm Rd; $21,000 loss
6/29	Accident on Farmington Falls Rd; David R. Dumont of Skowhegan; Fatality; Died in collision with Wolman Steel Company truck
6/30	Responded to 7 accidents, 2 vehicles, 2 false alarms, 2 fuel leaks, several misc. calls; Total 24 Calls for Month of June
7/02	Douglas Hiltz house (Chris Couller); Knowlton Corner Rd; $40,000 loss; Assisted by Industry, Temple, Chesterville & New Sharon
7/11	Accident; Car fire; Titcomb Hill; Christopher Keene car struck Angus cow in road; (Tracy's); Keene & Barbara Bean personal injuries; cow destroyed
7/18	MTE, Manufacturer Electronic Equipment, (Town of Farmington owner); Fairbanks; 8 hours; Total loss; $900,000; Assisted by Strong, Industry, Jay, Livermore Falls, New Vineyard, Wilton, New Sharon, & Chesterville; Worst fire in years
7/23	Carquest; Front Street; Light ballast in warehouse; Minor loss
7/27	Accident pick-up & bicycle; Hill Street; Anthony Adams, driving E. L. Vining truck, hit Barsbold Amarsaikhan on bicycle when crossed in front truck; Serious personal injury
7/31	Responded to 8 accidents, 4 false alarms, 1 stand-by for Wilton, 3 smoke calls, 3 MTE investigations, several misc. calls; Total 24 Calls for Month of July
8/10	Keith Howard property; RT 27 North; Grass fire; Discarded ashes
8/27	Search & Rescue: Youth lost; Searched from West Farmington to Wilton; Youth later found in Jay
8/31	Responded to 2 smoke calls, 5 misc. calls; Total 7 Calls for Month of August
9/02	Holly Lindbor house (Dana Goldsmith); Davis St, West Farmington; Structure fire; Suspicious; Minor loss
9/10	Rescue at Tumbledown Mountain; Loop Trail, Weld; Rescue & carry female who had fallen several feet down mountain cliff; Assisted Warden Service, Life Star Ambulance & Brunswick Naval Air Station helicopter with victim transport to Franklin Memorial Hospital

1998 MTE - Manufacturers Electronic Equipment Fire
Fairbanks School
Total Loss, $900,000

9/30	Responded to 4 accidents, 4 false alarms, 7 misc. calls; Total 15 Calls for Month of September
10/15	Search & Rescue; Washington Plantation, Wilton; Assisted Wilton to find Tom Baker; Suicide
10/31	Responded to 4 accidents, 3 brush fires, 2 service calls, 6 misc. calls; Total 15 Calls for Month of October
11/03	Butterfly Boutique (Jennifer Moes) Old Emery Store; 45 Broadway; Damage mostly to contents; $200,000 loss; Assisted by Industry, Wilton, Chesterville & Temple
11/05	Haz-Mat Call to Phillips Post Office; Pungent odor; Did not locate cause
11/09	Franklin Memorial Hospital; Board Room; Transformer for audio visual burned out
11/15	Krause Wood Products (Philip Krause); Depot St, Wilton; Boiler room; Damage to sawdust bin; Assisted Wilton
11/24	Darin Chase grass fire; Knowlton Corner Rd; Dumped ashes; 1 acre burned
11/30	Responded to 2 structure fires, 1 grass fire, 10 misc. calls; Total 13 Calls for Month of November
12/04	Joel Batzel house; 1 Bridge St; Suspicious; Minor damage
12/25	Accident; 2 cars by Irving Big Stop, RT 2 & 4; Gwendolyn Coffin; Fatality
12/31	Mutual Aid to Wilton; Then to Livermore Falls; Engine 6 & Ladder truck dispatched to fire at the Flower Barn & Apartments owned by Doug & Priscilla Mosher; Building total loss; Temperature 13 degrees
12/31	Responded to 10 accidents, 3 Mutual Aid, 1 structure, several misc. calls; Total 25 Calls for Month of December

1999	1/07	Accident; 2 vehicles, RT 2 & 4, Family Fare (Harvest House); Minor injury
	1/09	Accident; 2 vehicles head-on; RT 27, New Vineyard Rd; Fatality; Nina Bitterauf, killed; 2 other passengers injured
	1/18	Bill Groder property; Depot Street, Wilton; Assisted Wilton
	1/23	Accident, Snowmobile; Field behind Narrow Gauge Cinema; 1 man injured
	1/23	Accident; 3 vehicles with injuries; In front Old Big Apple, Lower Main St.
	1/23	Accident; 2 vehicles, icy roads; RT 27 New Vineyard Rd; Brent Demshire; No Personal Injury
	1/27	Webber Oil L. P. Gas leak; High Street & Stanley Ave
	1/31	Responded to 11 accidents, 4 extrications, 3 chimney fires, 1 Mutual Aid, several misc. calls; Total of 25 Calls for Month of January
	2/26	Arthur & Penny Brackett home; Starling St, West Farmington; Total loss; $40,000
	2/30	Responded to 4 accidents, 1 structure fire, 1 Life Flight, 7 misc. calls; Total 13 Calls for Month of February
	3/04	Arlo Henning's home; Back Falls Rd; Chimney fire
	3/08	Accident; Chip truck roll-over (Morse Bros.); RT 4 & 27; No injuries
	3/16	Accident; 2 vehicles head-on (Ackley, 32); RT 27 New Vineyard; Code
	3/17	Titcomb Hill Farm (Robert Mallett); Kitchen cabinet; Cause, toaster
	3/26	Accident; Car over bank; RT 4 Devil's Elbow; Irene Dodge; Extrication; Minor injuries
	3/30	Skowhegan Fair Grandstand, Para-Mutual Complex & Constitution Hall Buildings incl. 40 boats, campers & vehicles destroyed; $3 Million loss; 2:30 AM; 11 Fire Depts. called for Mutual Aid incl. Farmington.
	3/31	Responded to 3 false alarms, 1 Mutual Aid at 2:30 AM, 5 accidents, 2 spills, 12 misc. calls; Total 23 Calls for Month of March
	4/21	Harnden's Orchard grass fire; Bryant Rd, East Wilton; Assisted Wilton
	4/30	Responded to 2 accidents, 1 vehicle, 1 chimney fire, 1 Life Flight landing, 7 misc. calls; Total 14 Calls for Month of April
	5/12	Landfill Mulch Pile fire; Assisted by Chesterville & New Sharon
	5/31	Responded to 3 accidents, 1 Mutual Aid (Wilton stand-by), 10 misc. calls; Total of 14 Calls for Month of May
	6/03	Abbott House (MBNA owner); 121 Main Street; Total loss; 2 men arrested for starting fire; Arson; Assisted by Wilton
	6/16	Eugene Lambert storage shed; Farmington Falls; Destroyed; Total loss
	6/21	Ken Thompson property; Starks Rd, New Sharon; Minor loss; Assisted New Sharon

6/30	Responded to 6 accidents (1 with extrication), 4 vehicles, 21 misc. calls; Total 31 Calls for Month of June	
7/01	Police Department Assist; Amber Pond; Homicide	
7/31	Responded to 4 accidents, 3 hazardous spills, 1 Life Flight, 11 misc. calls; Total 19 Calls for Month of July	
8/02	Suzanne Lewis brush fire; Perham Hill Rd; Children playing with matches	
8/24	Mt. Blue Middle School; Middle Street; Gym floor damage; Spark from welding bleachers caught canvas covering floor on fire; $10,000 loss; Assisted by Industry & Wilton	
8/31	Accident needing extrication; Back Falls Rd; J. Kerr; Minor personal injury	
8/31	Responded to 9 false alarms, 2 vehicles, 18 misc. calls; Total 29 Calls for Month of August	
9/30	Responded to 26 misc. calls; Total 26 Calls for Month of September	
10/11	Milton Sinclair's barn; New Sharon; Minor Damage; Assisted New Sharon	
10/20	Vacant barn; Bubier Rd, Wilton; Total loss; Assisted Wilton	
10/31	Responded to 3 accidents, 4 smoke, 3 Haz-Mat, 2 Mutual Aid, 4 misc. calls; Total 16 Calls for Month of October	
11/13	Accident; Double vehicle roll-over; Paul Twitchell's Oil Truck & Harold Hargrave's Gravel Truck; Both trucks total loss; 900 gallons oil spilled;Beth Morris, personal injury	
11/25	Justin Serbo home; Industry; Interior damage; Assisted Industry	
11/30	Responded to 9 accidents, 3 trash fires, 3 spills, 2 Mutual Aid, 8 misc. calls; Total 25 Calls for Month of November	
12/07	Evacuation of Residents on RT 43 to Fairview Ave & Mt. Blue Middle School students on Middle to safe location; Bomb Threat	
12/07	Accident by Old Big Apple; Personal Injury; Used Jaws	
12/12	Accident CMP Worker; Brent Churchill, age 30; Fatality; Electrocuted repairing high tension line; Industry; Assisted Industry with Ladder Truck	
12/31	Responded to 6 extrications, 1 bomb scare, 1 accident, 1 electrocution, 2 service calls, 1 false alarm, 5 misc. calls; Total 17 Calls for Month of December	
2000	1/04	Robert Austin (Mt. Vernon) Apartments; 5 Thomas McClellan Rd; West Farmington; 3 Apts., 1st Apt., Shawn Austin, 2nd Apt., Christopher Neal & Rachel Cormier, Barn Apt., Sherry Tibbetts; Total loss; $85,000; Assisted by Temple, Industry & Wilton
	1/14	Raymond Pillsbury mobile home; 63 Cascade Leisure Park; Thawing pipes under home ignited insulation; $10,000 loss; Assisted by Wilton
	1/27	Roger Knudson; Beck Rd, Vienna; Total loss; 3 hours; Assisted Vienna
	1/31	Responded to 2 structures, 10 accidents, 1 extrication, 4 chimney, 3 training, 3 Mutual Aid (2 Wilton, stand-by, 1 Industry, accident), 12 misc. calls;

	Total 35 Calls for Month of January
2/11	Accident pick-up & tractor trailer; New Vineyard Rd; Michael Crocker, Jay; Fatality
2/13	Charles Fitz wall fire; 8 Starks Rd, New Sharon; Assisted New Sharon
2/15	Vincent Harris property; Intervale Rd; Temple; Assisted Temple
2/16	Farmington Chip Plant (E. J. Carrier Corp.); Town Farm Rd; Welding spark ignited wall insulation and studding; $25,000 loss; Assisted by Temple, Industry, New Sharon, Wilton, Strong & East Dixfield
2/17	Vance Child home and museum; RT 2, Dixfield; Total loss; Assisted Wilton
2/28	Responded to 1 structure, 3 vehicle, 2 CO Monitor, 13 accidents, 14 misc. calls, 3 Mutual Aid; Total 36 Calls for Month of February
3/01	Keith Howard Garage (Reginald Hanson); RT 27, New Vineyard Rd; 1985 F-250 truck owned by Gale Ellis burned; Caused by workman using cutting torch on truck body; $2000 loss
3/31	Responded to 2 structures, 4 vehicles, 8 accidents, 3 smoke, 3 CO Monitor, 8 misc. calls; Total 28 Calls for Month of March
4/02	Percy Harris Garage; Lane Rd, New Sharon; $80,000 loss; Assisted New Sharon
4/14	Sherwood Swain; Swain Rd, Industry; Basement woodstove flashed back; Burns on Swain's hands, arms & face; Damage to property $40,000
4/17	Maxell Road fire; Wilton; Minor damage; Assisted Wilton
4/30	Responded to 3 structures, 7 accidents, 1 vehicle, 3 false alarms, 2 medical alerts, 1 grass, 4 misc. calls; Total 20 Calls for Month of April
5/01-31	Responded to 3 accidents, 1 extrication, 2 Haz-Mat, 9 misc. calls; Total 15 Calls for Month of May
6/01-30	Responded to 1 Mutual Aid structure fire (Wilton); 16 misc. calls; Total of 16 Calls for Month of June. Total of 150 Calls for 1st 6 Mo.
7/01 - 12/31	2nd 6 Mo 105 Calls

Total of 255 Calls for Yr. 2000; 250 Emergency & 5 Service calls. (Structure, vehicle, accident, extrication, alarm malfunctions, Haz-Mat, gasoline & oil spills)

Chapter 9
FARMINGTON FIRE DEPARTMENT FIRE CHIEFS

1860	Captain Jennings	
1873	Captain F. V. Stewart	
1873 - 1886	Captain E. I. Merrill	
1886	George C. Purington	
1900 - 1904	J. M. Matheau	
1904 - 1908	George Blake	
1908	E. E. Flood	
1909	Arthur Tucker	
1910 - 1911	E. B. Kempton	
1912 - 1915	A. B. Carr	
1916	A. D. Keith	
1917 - 1919	E. D. Jackson	
1920 - 1922	Geo. Harry Koch	
1923 - 1951	Victor C. Huart	(28 yrs.)
1951 - 1969	J. Bauer Small	(18 yrs.)
1969 - 1977	Forest L. Allen	(8 yrs.)
1977 - 2000	Robert L. McCleery	(23 yrs)
2000 - Present	Terry S. Bell, Sr.	

Chapter 10

OFFICERS OF FARMINGTON FIRE DEPARTMENT
1930-2000

Farmington Fire Company No. 1

1930 - 1932 (First account of officers in order of rank)

Chief	Victor C. Huart
1st Asst.	Charles G. Nickerson
2nd Asst.	Arthur Locke
Treas.	Charles G. Nickerson
Clerk	Elden G. Hall

1932 - 1934

Chief	Victor C. Huart
1st Asst.	Chas. G. Nickerson
2nd Asst.	Arthur Locke
Treas.	Chas. G. Nickerson
Clerk	Elden G. Hall

1935 - 1942

Chief	V. C. Huart
1st Asst.	C. G. Nickerson
2nd Asst.	Arthur Locke
Treas./Clerk	C. G. Nickerson

1943 - 1946

Chief	V. C. Huart
1st Asst.	Roy Stinchfield
2nd Asst.	Elvet Gray
Sec.-Treas.	J. Bauer Small

1947 - 1948

1st Asst.	Elvet Grray
2nd Asst.	J. Ambrose Compton
Sec.-Treas.	J. B. Small

1949 - 1951

Chief	V. C. Huart
1st Asst.	J. A. Compton
2nd Asst.	J. B. Small
Sec.-Treas.	J. B. Small

March 27, 1951, Victor C. Huart Dies.

April 2, 1951, J. Ambrose Compton Elected Chief & Resigns.

April 2, 1951, J. Bauer Small Elected Chief.

1951 - 1956

Chief	J. Bauer Small
1st Asst.	J. Ambrose Compton
2nd Asst.	Delmar Johnson
Sec.-Treas.	Duane Hardy

1957

Chief	J. B. Small
1st Asst.	D. Johnson
2nd Asst.	Maurice Taylor
Sec.-Treas.	Duane Hardy

1958 (Sec.-Treas. only change)

Sec.-Treas.	Arthur Cutler

1959 - 1967

Chief	J. B. Small
1st Asst.	Maurice Taylor
2nd Asst.	Forest L. Allen
Sec.-Treas.	Arthur Cutler

Farmington Fire Department Officers 1959 - 1967

L-R: Arthur Cutler, Secretary- Treasurer, J. Bauer Small, Chief,
Forest Allen, Second Assistant Chief
Not Pictured: Maurice Taylor, First Assistant Chief

Courtesy Town of Farmington

1968

Chief	J. B. Small
1st Asst.	Forest Allen
2nd Asst.	Harrison Bragdon
Sec.-Treas.	Arthur Cutler

1969

Sec.-Treas.	Clifford Neil
Chief	J. Bauer Small Retires 12/02/69

1970 - 1974

Chief	Forest Allen (Elected Chief 12/2/1969)
1st Asst.	Harrison Bragdon
2nd Asst.	George Hobbs
Sec.-Treas.	Clifford Neil

1975   (Sec.-Treas. only change)

Sec.-Treas.	Robert L. McCleery

1975 - 1977

Chief	Forest Allen
Deputy Chief	Glenwood Farmer
2nd Asst.	George Hobbs
Sec.-Treas.	Robert L. McCleery

December 15 1976, 1st Asst. Chief, Harrison Bragdon, Dies.
March 31. 1977, Forest Allen Retires as Chief.
April 1, 1977. Robert L. McCleery Elected Chief.

1977 - 1983

Chief	Robert L. McCleery
Deputy Chief	Glenwood Farmer
Asst. Chief	George Hobbs, Retires 1/3/78
Asst. Chief	S. Clyde Ross
Sec.-Treas.	Richard Russell

1976 Farmington Fire Department Officers
L-R: Chief Forest Allen, Asst. Chief Norman Collins,
Asst. Chief Robert McCleery, Deputy Chief Glenwood Farmer, Asst. Chief Richard Russell

1977 Farmington Fire Department Officers
1st Row L-R: C. Blaisdell, R. Russell, C. Meader.
2nd Row L-R: Chief McCleery, C. Ross, G. Farmer, N. Collins, J. Bell

1983 Farmington Fire Department Officers with Chief Robert McCleery.

1985 Officers with Chief Robert McCleery
L-R: Asst. Chief Clyde Ross, Asst. Chief Terry Bell, Deputy Chief Glen Farmer, Asst. Chief Norman Collins

1984 - 1991

Chief	R. L. McCleery
Deputy Chief	Glen Farmer
Asst. Chief	S. Clyde Ross
Asst. Chief & Sec.-Treas.	Terry Bell
Asst. Chief	Timothy Hardy

1992 - 1st Half 2000

Chief	R. L. McCleery
Deputy Chief	S. Clyde Ross
Deputy Chief	Terry Bell
Asst. Chief	Tim Hardy

Effective July 1, 2000, Chief Robert L. McCleery retires. Terry S. Bell, Sr. becomes the town's first full-time Fire Chief.

2nd Half 2000

Chief	Terry S. Bell, Sr.
Deputy Chief	S. Clyde Ross
Deputy Chief	Tim Hardy
Senior Chief	John Bell
Senior Chief	Morrill Collins
Senior Chief	Harold Hemingway

On June 2, 2000, it was announced that Terry S. Bell, Sr. would become the town's first full-time Fire Chief. The town appropriated a salary of $33,500 with an additional $1000 after a six month probationary period plus a vehicle (two-ton pick-up truck.) According to June 16, 2000, *Franklin Journal*: "Words fly over Chief's salary" at the Board of Selectmen's meeting. Retiring Fire Chief Robert McCleery, other firefighters, Selectman Stephen Bunker, and Town Manager Pam Corrigan had recommended a starting salary of $36,000. Chief McCleery urged the board to either pay him what they should or don't have a full-time Chief. One of the selectman started the discussion with a proposed starting salary of $32,000 - the lower end of the scale - which firefighter, Clyde Ross, labeled as ridiculous and disheartening, stressing the nature of the Chief's position which includes being on call 24 hours a day, duties as emergency preparedness director and town fire warden. Stephen Bunker, a fire fighter and member of the board, citing potential appearance of conflict of interest stated the high end was a bit low to begin with and I am not shy about starting a candidate at the top - the job commands it."

1999 Officers
L-R: Captain Jack Bell, Deputy Chief Clyde Ross, Chief Robert McCleery, Asst. Chief Tim Hardy,
Deputy Chief Terry Bell, Captain Harold "Stub" Hemingway

Chapter 11

JR. OFFICER APPOINTMENTS
1978 - 2000

1978	Harold Hemingway	Promoted	Captain
	Jack Bell	Promoted	Captain
	Clyde Meader	Promoted	Lieutenant
	Clinton Blasidell	Promoted	Lieutenant
1980	Kenneth Durrell	Promoted	Lieutenant
	Morrill Collins	Promoted	Lieutenant
1984	Tim Hardy	Promoted	Lieutenant
	Nelson Collins	Promoted	Lieutenant
1988	Richard Knight	Promoted	Lieutenant
1991	S. Clyde Ross	Promoted	Deputy Chief
	Terry Bell	Promoted	Deputy Chief
	Tim Hardy	Promoted	Asst. Chief
	George Barker	Promoted	Lieutenant
1993	Morrill Collins	Promoted	Captain
1994	Stanley Wheeler	Installed	Chaplain
1996	Mike Bell	Promoted	Lieutenant
2000	Terry Bell	Promoted	Chief
	Tim Hardy	Promoted	Deputy Chief
	John Bell	Promoted	Senior Captain
	Morrill Collins	Promoted	Senior Captain
	Harold Hemingway	Promoted	Senior Captain
	Mike Bell	Promoted	Captain
	Nelson Collins	Promoted	Captain
	Richard Knight	Promoted	Captain

1916 Model T Ford Chemical Truck in Farmington
for 51st Maine State Firefighters Convention, Sept. 5-7, 2014
Photo Courtesy Owls Head Transportation Museum, Owls Head, ME

Chapter 12

TOWN MANAGERS
1974 - PRESENT

1974-1978	Philip K. Schenck, Jr.
1978 - 1986	Alan Gove
1986 - 1988	Dana E. Bradley
1989 - 1993	John G. Edgerly
1993 - 1996	Alphonse "Al" R. Dixon
1997 - 2000	Pamela S. Corrigan
2001 - Present	Richard P. Davis

Richard P. Davis
2018 Linc Stackpole Manager of Year Award
Maine Town, City & County Management Association

Philip K. Schenck, Jr.

Alan Gove

Dana E. Bradley

John G. Edgerly

Alphonse "Al" R. Dixon

Pamela S. Corrigan

Chapter 13

NARRATIVE

The original purpose of the fire department was to provide fire protection to the residents of the Farmington Village Corporation. The members of the fire department lived and worked within the Village Corporation limits. The boundaries of the Village Corporation were, and still are I believe, from the Sandy River to Tannery Brook (below South Street) to Middle Street up to the old Verne Millett slaughter house (stone wall) to Powder House Hill, and Box Shop Hill down Front Street to the river. As the town grew, the membership roster was expanded to include the Suburban Water District (Franklin Avenue).

In the early years of the Farmington Fire Department, there was no compensation for the firemen. Sometime in the 1900's, the Village Corporation approved a sum of money each year to be divided among all the firemen at the end of the year for fighting fires. This continued until 1960 when the town took over the fire department. For fires outside the Village Corporation hydrant district the firemen were paid by the town in which the fire occurred. The first record of any compensation to an officer of the fire department was in the 1930's. The Chief received $100, and the Clerk and Treasurer received $25 for the year. In 1939, the fire department payroll was $600.00 to be divided among the chief, janitor, and clerk & treasurer. In 1962, the fire department payroll was $2319.25.

In the 1950's, under Chief Bauer Small, some may remember the fire department holding an Annual Christmas Party for children up to 5th Grade. This was held at the old station on High Street with Santa, presents, and candy canes. This was fun for the firemen and the town's children.

From the early 1900's to 1970 there was an active fire company in West Farmington which was formed to provide fire protection in that part of town. They responded to all fires in West Farmington and assessed West Farmington residents a fee for fire support. Fund raising events were suppers and socials; they were famous for their oyster stews. The Town of Farmington made small contributions to the West Farmington Fire Department for their assistance on major structure fires within the Village Corporation. To this day [2000] the West Farmington Social Club still holds meetings on the first Thursday of the month. (Several members of the West Farmington Fire Department joined the Farmington Fire Department when the companies merged in 1970.)

There was an active fire company in Farmington Falls until the Farmington Falls department was accepted as part of the Farmington Fire Department at the annual town meeting on March 2, 1976. A sub-station is maintained at the Falls with an engine housed there to provide coverage to the lower part of town. The Falls firemen sponsor many events and activities for young people. They have an annual field day with parade and barbecue, an auction, and hunter's breakfast to raise money for such things as the playground, little league ballfield, and athletic equipment. They hold many benefit suppers for the elderly and citizens in need. (Note: The Farmington Falls sub-station is no longer manned, but equipment is stored there.)

In 1977, I began restructuring the fire department into a chain of command system; the department went from Chief and two (2) Assistant Chiefs to Chief, Deputy Chief, Captain, Lieutenant, and Private. This has served as an incentive to promotion and advancement, and has allowed me to recognize the contributions of more of our dedicated, hard-working fire fighters.

As the town has grown, fire equipment and fire-fighting techniques have changed significantly over the years. I will list some of the changes that have occurred in the operation of the fire department since 1977. The cost of replacing fire apparatus has gone up an ever-increasing amount, until replacement costs for town apparatus today has reached over $300,000 in 2000. (Note: In 2007, the Pierce 100 foot Ladder Truck "Tower #3" cost over $800,000.)

Fire training and education requirements and mandates are constantly changing. The use of protective breathing equipment requires extensive training and certification before a fireman can enter a hazardous area. The Department of Labor Occupational Safety and Health Association (OSHA) annually checks the training records on breathing equipment, medical records, and turnout gear. The requirements for turnout gear have changed more in the last 20 years than anything else going from the rubber raincoat to the Duct type canvas coat to a fire-retardant material with vapor barriers (Nomex-Kevlar barrier-type gear). A fireman today is dressed in a short coat and pant suit with short safety boots. The old hip boots are gone. Today's fireman wears an approved helmet, hood, safety glasses, Nomex-Kevlar type fire retardant pant suit, fire resistant gloves, and safety boots. The cost to outfit each fireman is $1200 to $2000. To train a fireman to Firefighter II Certification today costs $3000 and up. For those firemen who train to the Hazardous Materials Technician level there is no end to the training and cost.

This area is fortunate to have had International Paper Company's interest and expertise in funding and training a Haz-Mat Team. They have furnished and equipped a Haz-Mat vehicle that is available to area towns under a Mutual Aid agreement. There are many tank trucks and box trailers hauling hazardous material and waste through our area every day and night. This constitutes an accident waiting to happen and this area is prepared. August 21, 1990 is a case in point when an industrial chemical waste spill occurred with all mutual aid companies responding as did the International Paper Co., Environmental Protection Agency, and the transport company's own Haz- Mat Team.

Over the years, fire hose has changed from 2 ½ inch attack and supply lines to 1 ½ - 1 ¾ inch attack lines, and from straight tip to adjustable gallon nozzles, and to Task Force tip automatic nozzles. Large diameter supply lines, 4 inches, are a part of our standard operating procedures. The fire department at present has 25 air packs with a spare bottle for each. Portable folding water tanks are a part of our rural fire practices when there is no water available. This department has three (3) 2100 gallon folding tanks carried on the engines. The fire department has generators of different sizes, and an assortment of lighting equipment available for most any emergency. We also have a portable 500 GPM (gallon per minute) pump for off highway use and at forest or woods fires.

The Farmington Firemen's Benevolence Association is comprised of members of the fire department with their own treasury, separate from the town. They raise money from barbecues, the Farmington Fair booth, Maine State Firefighter's Conventions in 1985 and 1994, and other special events. They have donated money to charitable causes as well as purchasing several pieces of equipment to be added to the fire department at no cost to the town. Some of the equipment purchased has been: Poseidon air compressor; 6 cascade bottle system to fill air bottles made possible by supplemental donations from our mutual aid departments; 4 cascade bottle system added to the Squad Truck; 2 hand generators; and a portable folding tank (2100 gallon capacity).

In 1992, Commercial Rich Services, Inc. (ISO) conducted a Fire Insurance Classification rating of the Fire Suppression Services of the Town of Farmington. This consists of an evaluation of apparatus testing records, training records, engine company manning, dispatch procedures, and water supply (hydrant flow testing). The Town of Farmington has a Public Protection Class V. This is in a range or class from one (I) to ten (X). This rating pertains more to commercial buildings, but it does impact insurance rates.

The Farmington Fire Department has a written mutual aid agreement with our neighboring town fire departments. No town or city can exist today without this commitment. In 1990, written mutual aid agreements were signed by town officials and the Fire Chief, and today these agreements have become standard operating procedure due to liability issues. The fire department responds to many, in and out of town, accidents and extrication calls, and calls to aid the Police Department with traffic control, so standard operating procedures and mutual aid agreements are necessary to protect the town and individual.

The Fire Department has acquired important equipment for enhanced Search and Rescue. In 1998 a new Hurst tool with rams, cutters, and spreaders (Jaws of Life) was added to the rescue equipment, replacing the 15 years old air-operated Porter Power Tool. The Jaws of Life are hydraulic rescue tools powered by a piston-rod hydraulic pump. Our other search and rescue equipment includes the following: safety ropes of various sizes, pulley and hardware for high and low angle rescue, rescue rocket gun to throw a line over a river or gully, cold water survival suits for river or lake rescue, and ice and snow rescue sled. We maintain a complete record of all firemen who have satisfactorily completed the training requirements for using all safety equipment.

During the time of Vic Huart and Bauer Small's tenure, Florence Norton King immediately went to the fire station when the whistle blew. She was the Ladies Auxiliary before the Ladies Auxiliary existed. She prepared sandwiches, lots of hot coffee and donuts for the returning firemen. If there was a large fire, she would prepare soups. Many a frozen fireman was thawed by her generosity.

In the early 1980's, a group of firemen's wives and friends organized the Ladies Auxiliary for the purpose of furnishing food and beverages at fires or emergencies when needed. The ladies hold meetings and special events to raise money to support their programs. The Ladies Auxiliary of this department, and those of our neighboring fire departments, work together at many fire and emergency scenes. The Farmington Firemen's Ladies Auxiliary deserves our never-ending gratitude for being there to support "their men."

From 1990's to present [2000], the Firemen's Benevolent Association and the Ladies Auxiliary have worked together on the annual Easter Egg Hunt for the boys and girls in town, ten (10) years old and younger. They place 100 eggs about town and provide clues to find them. This creates a lot of interest. Each egg contains a slip noting the prize won. There are also grand prizes, usually bicycles, presented to a couple off winners.

In 1912, you had to live within the Village Corporation limits to become a fireman so as to hear the fire alert system. This was the case until 1960. Farmington fire department members according to the 1912 Bylaws had to be of "good moral character and temperate habits". Any member who appeared visibly intoxicated at a fire, or at the station, was to be reported to the Finance Committee, and if after an investigation, found guilty, would be dismissed from the Fire Company. There is no record of anyone having been dismissed for that reason. Failure to attend meetings for a period of 3 months without an excused absence was cause for being dropped from the membership rolls; any member who was "guilty of insubordination, inefficiency, or neglect of duty" was also dropped from membership. Today to be a member of the Farmington Fire and Rescue Department a person must be at least 18 years of age and have a high school education. Failure to attend meetings, insubordination, and inability to carry out one's duties are still reasons today to be dropped from the department. (See Chapter 16 - Charter & By-Laws)

Article 2 of the 1912 By-Laws state, and I quote: "The object of the company shall be to care for and operate the Fire Department of the Farmington Village Corporation for the purpose of extinguishing fires and the protection of property in the town of Farmington from loss by fire." Today the written statement of purpose of the Farmington Fire Department states: "It shall be the purpose of the Farmington Fire Department to provide fire and rescue protection and such other emergency services as the members are trained and equipped to perform. To answer such calls in the vicinity of Farmington as directed by the Administration of the Town of Farmington and the Fire Chief. The Fire Department shall honor the written and mutual agreements with other neighboring communities."

The Chief Officers of the Farmington Fire Department have been involved in many outside activities, past and present. J. Bauer Small was involved in organizing the Franklin County Firemen's Association, and its first president. He was a member of the Maine Fire Chief's Association, and the New England Association of Fire Chiefs. I have served as Franklin County vice president of the Maine State Federation of Fire Fighters, and president of the Franklin County Firemen's Association. I served on the board of directors of the Maine Fire Chief's Association for thirteen (13) years, and president of the State Association for one (1) year. The Fire Chief's officers and chaplain are also members of the Maine Fire Chief's Association. Deputy Chief S. Clyde Ross and I are members of the International Association of Fire Chiefs, New England Division, and the New England Association of Fire Chiefs.

In the past 25 years the Fire Department budget has gone from $10,000 in 1972 to approximately $225,000 in 1999. This is due to the increase in the cost of protective clothing, public safety equipment, and wages. It will continue to increase dramatically until such time as the town employs full-time firemen.

From its very existence, the Farmington Fire Department has been served by dedicated Chiefs and Chief Officers. Chief Victor Huart and those before him served at a time when they did not have fire equipment available to attack large fires. With the purchase of the Maxim Pumper (500 GPM) in 1931 Chief Huart's fire-fighting capability improved, but if there was no water available to draft from, all the department could do was stand-by while the building(s) burned. After the 1947 forest fires, a 750 gallon tank truck was purchased to complement the Maxim Pumper. Chief Bauer Small was instrumental in up-grading the department's fire-fighting capabilities by adding two (2) pumper tankers to the fleet in 1954 and 1966. Mutual Aid was called only as a last resort in those days and usually too late. Under Chief Forest Allen the department continued to improve with the purchase of a Mack Pumper (1000 GPM) with 750 gallons of water. Now the fire department has three (3) trucks capable of carrying 2500 gallons of water to the scene, a Tanker Truck, 110 foot Ladder Truck, a Squad Truck with winch, portable water pump, and Honda generator for fire scene lighting, and enhanced search and rescue equipment that includes extrication tools such as the Jaws of Life.

Prior to June 26, 2000 there were no full-time firemen in Farmington or in any other town in Franklin County. This changed when my successor, Terry S. Bell, Sr., became the town's first full-time Fire Chief. I look forward to the day when our department is staffed with more full-time firemen.

In 1964 Chief Bauer Small stated in his annual report the following: "…eventually we will be forced to maintain permanent men at the fire station. This is not an immediate necessity, but as the town grows and the Underwriters demand more protection, it certainly will be a requirement… like all municipal departments, there is a constant demand for more and varied services." In 1977, the Insurance Services Office (ISO) conducted a rating survey of the fire department with recommendations for upgrading and improving the fire department. This included the need for a full-time chief and sufficient company officers. It was noted in my 1977 Annual Report that the need for some permanent firemen was not far away as it was continuously harder to summons daytime help for fires, and more demands were being made for chimney inspections, and building permits from the State Fire Marshal's Office for day care centers and foster homes. In 1978 and 1979, I again stated there was a need for some full-time firemen (maybe 3 at the station during the day) to answer emergency calls. In 1984, I stated the following: "As the town grows and expands in both residential and commercial areas, a greater demand is being placed on the Fire Department, and some consideration will have to be given to full-time firemen. Employers are becoming reluctant to release their help to respond to fire calls, some will not do so even now. There is a need for inspections by the local Fire Department of new construction, remodeling of apartment and rooming houses, day care centers, homes, businesses and places of amusement." In 1986, 1987, and 1988, I reported that the town needed some full-time firemen to adequately perform needed building inspections, apparatus and equipment maintenance, pre-planning, and record keeping required by the unfunded mandates of the State and OSHA. In 1994 and the years that followed, the requirements and responsibilities of the fire department were on the increase. Annual inspections from OSHA and the Department of Labor Standards required records on training, purchase of turnout gear, medical exams, S.C.B.A. maintenance, Hepatitis B vaccinations, as well as apparatus and equipment maintenance. Upon tendering my resignation in January 2000 effective July 1st, I made it clear that now was the time to transition to a full-time chief.

Note:

Change takes time; sometimes it takes 40 plus years. The year 2000 was a pivotal year for the Farmington Fire Department with the installation of its first full-time Fire Chief. Chief Bob McCleery did not live to see the establishment in 2007 of two (2) per diem fire fighter positions, and the number increased to three (3) in 2014, nor the town approve the hiring of four (4) full-time fire fighters in 2016, but he would have been most pleased for the town and Chief Bell. The full-time fire fighters hired were: Timothy D. Hardy, Scott Baxter, Joseph Hastings and Seth Abbott. In 2017, Chief Terry Bell had four (4) firefighters including himself positioned at the station on any given day.

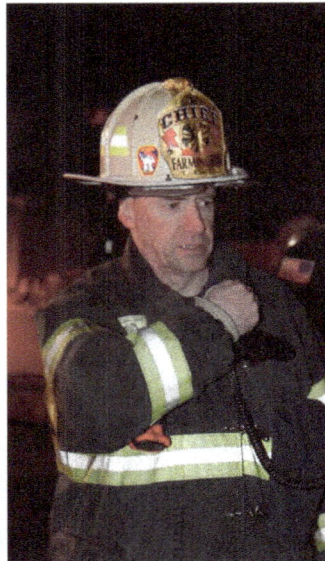

2014 RT Photo Collage: Farmington Firemen on the job;
Deputy Chief S. Clyde Ross with Chief Bell; Deputy Chief Tim Hardy; Chief Terry Bell.

LT Photo Collage: Chief Bell with Dep. Chief Ross, white helmets, & Fire Rescue crew;
Captain Mike Melville with Captain TD Hardy & firemen at Strong fire; Captain Mike Bell,
Jaws of Life, with Captain TD Hardy; Firemen of Tower 3, Mike Cote; Deputy Chief Tim Hardy.

Photo Collage Courtesy Town of Farmington

Chapter 14
LIGHTER MOMENTS IN THE FIRE DEPARTMENT

I will reminisce about my involvement with the fire department and some of the lighter, more humorous events that occurred in the 50 years that I have chased the red fire truck. Living two (2) miles from the fire station and outside the limits of the suburban water district, I could not belong to the Farmington Fire Department. The Chief said that I could not hear the alarm so as to make the trucks upon dispatch. Well, that was all the incentive I needed.

In those days, and until into the late 1960's, the fire alarm was sounded by an air horn (fog horn). On a clear day, or night, the sound would carry for miles…at least two anyway. The calls used were 1 blast - Call out a driver; 2 blasts - All out for fire; 2 blasts repeated in two minutes at 7:00 AM - No school; 2 blasts at 12 noon - Testing; 2 blasts at 7:00 PM first Tuesday of the month - Call out for a meeting; 4 blasts - Chief's call out for a fire; 5 blasts - Company call out for practice; 5 blasts repeated (55) - Company call out for an out of town fire. This was in the day of pull box alarm boxes at street corners around town which have since been disconnected due to upkeep costs and false alarms; today we have the 9-1-1 system.

When J. Bauer Small became Fire Chief in 1951, I again tried to join the Fire Department, but with the same result. It was not until 1960 when the town acquired the fire department from the Village Corporation that I was able to officially join. In the meantime, I had my own raincoat, helmet, and boots and upon hearing the horn would race to the station where I was allowed to ride on a fire truck and work at a fire scene. Sometimes I even beat in-town firemen to the truck which was a source of pride to me and amusing to the group. This is a great big, NO, NO today for liability reasons; times have changed.

From the beginning, the Fire Department has held regular oyster stew suppers and invited guests to attend. Firemen are known for their camaraderie, and firemen gatherings could get very interesting at times when it involved the partaking of the spirits. A popular event was the annual lobster cook-out held the first weekend of September. Two or three firemen would take the lobster box and head for the coast, pick-up lobsters, and be back in time to cook them for dinner, usually. This event was held at different places. The first one I attended was at Tom Clark's camp on Clearwater Pond with horse shoe pitching, softball games, and boat rides. What an experience! A few years later the event was moved to Powder House Hill in back of the reservoir under the pine trees. On a hot Sunday in September as the men were sitting down to eat their lobsters, an alarm came in for a fire at Armstrong's chicken barn on the Fairbanks Rd, now a car wash and electric supply business. When the firemen returned from the fire the lobsters and beer were gone. At the annual lobster cook-out there were firemen who were noted for making a special brew that was very popular at this event. In fact, it was so good that a few men were known to be sleeping under the trees after the party was over. Some were even known to go swimming late in the afternoon. I am sure you can guess where! The brew making came to an abrupt halt after a batch blew-up in the maker's basement. From then on, the lobster outing was held in the back of the High Street Station, at the Falls, and on the lawn of the treatment plant. This event has been replaced by our annual meeting supper held in the training room at the fire station and is a much quieter affair.

At times citizens complain about firemen over driving and speeding to the fire station. They have nothing on me. In responding to calls, I have driven cars off the road, been chased by a State Trooper, and missed the Abbott Hill turn ending up in the college parking lot one rainy night. After my experience with the State Trooper, the firemen received Red Lights for the front of their vehicles. This resulted in a lot of good humored ribbing as you might imagine. I have slowed down some over the years.

As you may hear tell, I should be a lot older than I am based on the number of birthday celebrations I've had with my good buddies over the years at conventions and outings. I should probably be twice my age if anyone were counting. It became quite a ritual to have the restaurant wait staffs sing "Happy Birthday" to me. The good stuff I did with the help of all in my department; the other stuff I did with the help of my good buddies. It's been one heck of a ride!

Chapter 15
THE FIRE ALARM SYSTEM

In the late 1800's and early 1900's, fire bells alerted firemen to report for a fire. Prior to that church bells most likely sounded. In 1912 an audible air horn was installed on the fire station, and fire alarm boxes were located throughout the Village Corporation limits.

On January 30, 1912, the Farmington Village Corporation adopted a Charter and By-Laws for Farmington Fire Company No. 1 which detailed fire alarm boxes and locations, instructions on use, and fire and special signals descriptions.

The following is an account taken from that document:

Special Notice.

For information regarding the Alarm System or the Fire Department call up A. B. Carr, Chief Engineer, Main 111-5, Farmers' 30-2; or A. D. Keith, First Assistant Foreman, Farmers' 20-23. If any defect is known to exist in the Alarm System, or if any tampering with it is known, parties are requested to notify the Chief Engineer as above.

Boxes and Locations.

There are now 12 fire alarm boxes within the limits of the Corporation. They are numbered and located as follow:

12 High Street, in front of Public Library.

13 Upper Main Street in front of Keith House.

14 East Corner of School and Perkins Street, opposite Louis Baker House.

21 Southeast Corner Broadway and High Street, near Keyes House.

23 South Corner Main and Park Streets, opposite Northeast Corner of Common.

24 Perham Street, Corner nearly opposite residence of H. B. Coolidge.

31 Perham Street, opposite foot of North Street.

32 Main Street, opposite South Street, in front of F. W. Butler's residence.

33 Southeast Corner Main Street and Broadway, in front of People's National Bank.

41 East Corner Court and Grove Streets, in front of Titcomb House.

42 Southeast Corner Middle and Quebec Streets, near Catholic Church.

43 West Corner Anson and High Street, in front of Dobbins House.

Boxes Outside the Corporation.

In putting in the present system provision was made for a few boxes outside the Corporation, and they may be connected with the System at any time the parties interested put in the boxes. These Assumed Boxes are designated by the following numbers:

15 Suburban Water District – Later Farmington Shoe vicinity.

22 West Farmington M.C.R.R. District.

44 Box Shop and Vicinity.

Instructions.

In case of fire go to the nearest Alarm Box. To give the alarm break the glass in the door, turn the key, open the door, then pull the hook to the bottom of the slot and let go. Do not pull the hook a second time, or allow anyone else to do it except an officer. Stay by the Alarm Box until Firemen arrive to direct then to the fire, and to see that no one interferes with the Box. (In case the alarm does not sound at once go to the next nearest Box, and sound the alarm from that.) Do not give alarm for a fire seen at a distance.

The Fire Signal.

When the Alarm is pulled at any Box, the number of blows or toots, will correspond with the figures on the Box from which it is given. Three short blows, followed by three longer blows come from Box 33, at the southeast corner of Main and Broadway; one short blow, followed by four longer blows comes from Box 14, and so on; These signals are repeated three times or sounded four times in all.

Special Signals.

Number 22, two blows, repeated twice at 7:45 or 11:45 AM, closes all public schools for one session.

Number 22, two blows, repeated once between the hours of 7:00 PM, and 5:00 AM, calls out the Police of the Corporation.

Two blows at 1:00 PM, gives the Standard Time.

Five blows repeated calls the Fire Company out for practice.

One "round" or the number of box sounded at 6:00 PM, on a Saturday indicates a test by the Chief Engineer.

The Alarm repeated from any Box constitutes a Second Alarm and calls out the Fire Companies from West Farmington and Suburban Water District.

Number 44 will call the Fire Company to the vicinity of the Box Shop, where there is no Box at the present time.

Fire Alarms from West Farmington and the Suburban Water District are made by special arrangement with the Maine Telephone Co.

Starting with the tenure of Chief Victor Huart and 1st Assistant Chief Chas. Nickerson the following were the Special Signals:

Two blasts repeated in 2 minutes at 7:30 AM and 11:30 AM. No School.

Two blasts mean "Fire all out."

Two blasts at 12 Noon, Standard Time.

Five blasts call the Company out for practice.

Five blasts repeated calls the Company out for outside fire.

Four blasts, Chief's call for fire.

Six blasts, Emergency call.

One blast, driver's call.

Two blasts at 7 PM first Tuesday of month. Firemen's Meeting.

Alarm repeated from any box constitutes a Second Alarm and calls out the fire companies from West Farmington and the Suburban Water District as well as Farmington.

During Chief Victor Huart's tenure which began in 1923, the fire alarm was sounded from a horn at the old fire barn (formerly CMP building), or from Victor Huart's Barber Shop on Broadway, or from a street box. After the new station was built on High Street (now Farmington Water Department) a new fire alarm console was purchased with the capacity of transmitting several types of alarms via radio frequency directly to firemen as well as the ability to switch on and off a siren. With the election of J. Bauer Small as chief, in 1951, the unit for blowing the horn was moved to W. W. Small Store on Lower Broadway and Pleasant Street. On June 3, 1958, more efficient audible alert horns were installed on the High Street Fire Station.

Starting in 1961 Chief Bauer Small began the process of shifting the department over to a Plectron "quick alert" radio system which could be activated from his house on High Street or the fire station. This system was tied to radios distributed to the firemen for first call response. In 1962, there were 5 radios and they were distributed to firemen on a six month rotational basis starting at the top of the membership roster.

Fireman Red Phones had been in use from the time telephones were available. These red phones were distributed to department officers to alert them to call out for fires. There was also a Red Phone at the fire station, Chief Bauer Small's store, and the Farmington Falls store on the corner. These phones received incoming calls only. Once a call to the Fire Department (778-2120) was received, Red Phones were called, the Plectron "quick alert" system activated with radios signaled, and the siren or horn blown. Upon the retirement of Chief Bauer Small and the election of Forest Allen as chief, the alarm unit and dispatching console went to the Allen's home on Court

Street. Mrs. Vertie Allen dispatched fire and emergency calls from her home between 1970 to 1983 at which time dispatching was transferred to the old County Jail, then moved to the new jail.

Fire and emergency calls were also dispatched many a night from the fire station when Frank Gagne who resided at the fire station, although not a fireman, received a call and activated the alert systems and horn. Once the Plectron "quick alert" radio system was fully implemented the audible horn was slowly phased out, and Red Phones were discontinued in 1985. All emergency dispatching today is done from the Franklin County Correctional Dispatch Center (Jail) which has twenty-four (24) hour coverage. (Today it's the Franklin County Regional Communications Center.) This center covers calls for most of the Franklin County Departments.

Chapter 16
1912 FARMINGTON VILLAGE CORPORATION CHARTER, BYLAWS & DEPARTMENT RULES AND REGULATIONS

Contributions of Farmington Fire Co. No. 1

Some of the rules and regulations you will read below are still in effect today. Some have been changed to meet the Department of Labor's safety requirements. Others have been deleted or changed due to restructuring of the fire department, and some have been modified to reflect the evolution of apparatus as well as fire-fighting techniques.

(Special thanks to Captain Harold "Stub" Hemingway who provided me with a copy of the Charter & By-Laws for this research.)

Taken from the Farmington Village Corporation Charter and By-Laws of 1912:

The CHARTER

Chapter 142, Private and Special Laws of 1911

Approved by the Governor, March 20, 1911. Adopted by the Farmington Village Corporation, January 18, 1912.

Section 2 Paragraph 4: To Maintain a Fire Department

To organize and maintain an efficient fire department, and to adopt all rules and regulations for governing the same.

The By-Laws

Adopted at the annual meeting held January 30, 1912

Article 2.

Officers to be Appointed by the Assessors.

In addition to the officers provided by the charter for election by ballot, the following officers shall be appointed by the Assessors and shall be by them removed for cause, viz.:

A collector of taxes whenever taxes are levied.

A collector of water rentals for the water department.

Chief of Police.

Chief engineer of the fire department.

Superintendent of the water department.

Night watchman -- one or more.

Four Fire Wardens.

The compensation of the above named officers, as well as those chosen by ballot, shall be fixed by a vote of the Corporation at its annual meeting or any special meeting called for that purpose, and each of said appointive officers shall be appointed by the Assessors, upon their qualifications, to serve until the next annual meeting or until their successors are chosen unless before the expiration of said term, they shall resign or be removed, when the vacancy so occasioned shall be filled by a new appointment.

Article 5. Chief Engineer of the Fire Department

The chief engineer of the fire department shall have control and management of the fire department, appliances and apparatus, and subject to the approval of the assessors, adopt rules and regulations for governing the same; select his subordinate officers and members of the fire company and have general charge of all matters pertaining to said department. He shall keep the apparatus of the department in perfect working order at all time, and keep the assessors informed of its condition.

Article 8. Four Fire Wardens

The assessors shall appoint two fire wardens whose duties shall be the same as those of the fire wardens of towns, and shall appoint two other fire wardens whose duties shall be, whenever requested by the assessors, to enter all the building of the corporation wherein fires are maintained, examine into the conditions as to the security against fire, and order such changes as may be necessary to protect persons and property from damages of fire.

Constitution of the Farmington Fire Co., No. 1

Article 1. Name

The name, style and title of this association shall be the Farmington Fire Company, No. 1

Article 2. Object

The object of this company shall be to care for and operate the Fire Department of the Farmington Village Corporation, for the purpose of extinguishing fires and the protection of property in the town of Farmington from loss by fires.

Article 3. Members

It shall consist of as many members as are necessary from time to time to operate the Fire Department. The members shall be residents of the Farmington Village Corporation, of good moral character and temperate habits, and may become members by complying with the By-Laws of the company.

Article 4. Officers

The officers shall consist of a Foreman who shall be the Chief Engineer of the department and shall be appointed by the Assessors, the First and Second Assistant Foreman, Clerk, Treasurer and a Finance Committee of three, one of whom shall be the First Assistant Foreman. These officers with the exceptions named and the Finance Committee shall be elected by ballot at the annual meeting held on the first Tuesday of December.

There shall also be the following officers who shall be appointed by the Foreman before the first of January following the annual meeting in December: Engineer and Assistants, Fireman, Foreman of Hook and Ladders, First and Second Assistants, two Pipemen [nozzle] for each line, two Assistant Pipemen for each line, two Hydrant men, two or more men to adjust suction-hose, and one Lineman who shall be appointed a Special Police by the Assessors.

Article 5. Meetings

Section 1. The company shall hold regular meetings on the first Tuesday of each month beginning at 7:30 PM and the December meeting shall be the Annual Meeting for the election of officers.

Section 2. Special meetings may be called by the Foreman, or by the Clerk on petition of three or more members of the company.

Article 6. Amendments

This Constitution may be amended by a vote of the company at any regular meeting, provided the proposed amendment has been submitted in writing at a previous meeting and laid on the table at least one month.

By-Laws of the Farmington Fire Company, No. 1

Article 1. Duties of Members

Section 1. A majority of the members present shall constitute a quorum.

Section 2. Should the Chief be absent the next officer in rank, (See Article 4 of the Constitution) shall preside.

Section 3. Candidates for membership must be proposed by a member, who shall set forth in writing the name, age, residence and occupation of the applicant, and the application shall be referred to the Finance Committee.

Section 4. No person shall be considered a member of the company until he has signed the Constitution and By-Laws.

Section 5. After a fire is extinguished and the Foreman having command considers the apparatus of no further use, the members of the company shall assist in taking up the hose and return with the apparatus to the engine house, the roll shall then be called, after which the members shall be at liberty to depart.

Section 6. If the alarm that causes the apparatus to be taken out should prove to be unfounded the members shall, nevertheless, consider themselves equally under the government of the Foreman, as though a fire had actually taken place.

Section 7. Any member who shall appear visibly intoxicated at a fire, or on or about the premises of the company or at any time when acting as a member of the company, shall be reported to the Finance Committee, and if after due trial and investigation, he be found guilty, he shall be discharged from the company.

Section 8. Neither the Constitution nor the By-Laws shall be altered or amended, unless the alteration or amendment be proposed in writing one month previous to adoption and consented to by two-thirds of the members present.

Section 9. Failure to attend meetings or to respond to fire alarms for a period of three months without an excuse deemed satisfactory and approved by the signatures of the Finance Committee, will be regarded as sufficient cause for dropping the member from the rolls of the company and the clerk shall notify the member.

Section 10. The engine house and hall shall be closed not later than 10 o'clock PM and shall be kept in good order by the janitor. No intoxicating liquors shall be allowed in or about the premises.

Section 11. A member shall be dropped from membership who is guilty of insubordination, inefficiency, or neglect of duty. (See Sec. 9.)

Section 12. The members of the company will be furnished with rubber coat, hat, boots and rubber mittens, which belong to the Department.

Article 2. Duties of Officers

Section 1. --Foreman.

1. The Foreman shall preside at all meetings of the company, preserve strict order, enforce decorum and the Constitution and By-Laws. He shall decide on all questions of order, unless an appeal is made to the company, inspect ballots for candidates for membership and report thereon to the company.

2. He shall have a casting vote on all questions before the company, having an equal number of votes.

3. He shall sign all orders and notices and appoint committees except when otherwise ordered.

4. He shall call special meetings whenever he thinks it necessary, or at the request of three members.

Section 2. --First and Second Assistant Foreman. It shall be the duty of the First and Second Assistant to aid their superior officer in his work as Foreman and in his absence to act respectively in order of rank in his place.

Section 3. --The Clerk.

1. The Clerk shall keep a correct record of all meetings of the company in a book for that purpose, which at all meetings of the company shall be open for the inspection of the members.

2. He shall at each meeting of the company read the records of the preceding meeting. He shall call the roll at precisely the time stated for the meeting of the company and also after the return from any fire or false alarm and shall record the names of the members present in the record book.

3. He shall notify the members of their admission to, or discharge from, the company, and also of their being appointed on any committee.

4. He shall notify all regular and special meetings, when directed to do so by the Foreman, by written

or printed notice, and shall keep a correct roll of all members and each member's absence from fires, and monthly, or special meetings.

Section 4. --Treasurer.

1. The Treasurer shall have charge of the funds of the company; keep and render to the company when called upon by them a true account of all moneys received and paid out by him.

2. At the annual meeting he shall render to the company a correct statement of the financial affairs of the company, and at the expiration of his office deliver to his successor all books, papers and moneys in his possession belonging to the company.

3. When directed to do so by the Finance Committee he shall pay all bills or orders approved by a two-thirds vote of the company and signed by the Foreman and attested by the Clerk.

Section 5. The Finance Committee.

1. At the beginning of each year there shall be appointed three members, one of whom shall be the First Assistant Foreman, to act as a Finance Committee.

2. They shall judge all fines and excuses for non-attendance at fires and meetings.

3. They shall examine and approve all bills against the company before they are paid, and examine the Clerk's and Treasurer's books whenever they deem it necessary.

4. They shall receive all applications for membership and shall report the candidate to the company, if approved by them, he may be balloted by ball ballot at any monthly meeting thereafter, and any candidate having not more than two black balls shall be declared elected. In case there in no vacancy at the time, if favorable action is taken, the name may be placed upon the waiting list, and then be received in order as vacancies may occur.

Rules and Regulations of the Fire Department of the Farmington Village Corporation

The Fire Company

In a general way the duties of the company are set forth in Article 2 of the Constitution. Orders from superiors must be promptly obeyed and every man must attend to the duty assigned him, and not leave his post until the duty is finished.

The members of the Fire Department whenever there is an alarm of fire shall repair promptly to the engine house for duty, where they shall assist, under the direction of the Second Assistant Foreman, in removing the apparatus to the scene of the fire, and then exert themselves in the most efficient manner possible in working and managing the apparatus, and in performing any duty that shall be required of them by the officers of the department. They shall avoid opening the doors and windows of any building where a fire may be, until a supply of water can be procured, apply it directly to the parts on fire, and carefully avoid damaging goods or furniture by water. They shall do everything possible to protect the property of the Department and after the fire join in returning the same to the engine house and then cleaning it and putting it in order for future use.

In case any break or other difficulty is discovered with the apparatus, it should be reported to the Chief at once.

It shall also be the duty of all to aid the officers and obey their orders to the end that the company may always be ready and efficient firefighters. There are often instances where there is danger from fire and attention called to them will prevent much loss of property. It should be just as much a duty to prevent fire as to extinguish one.

Special Duties of Officers and Others

Chief Engineer or Foreman. When an alarm of fire is sounded it shall be the duty of the Chief to repair at once to the locality indicated by the call. To examine the situation of the fire and determine what work shall be done by the company as soon as it may appear, to direct where the hose shall be attached to hydrants and pipe lines laid, ladders placed and streams directed upon the fire, and attend to such other duties as may appear necessary. He shall at all times preserve order at and around the fire, watch closely the progress of the fire, and place the firemen where they can work to the best advantage in extinguishing the flames.

After the fire is extinguished he shall return the apparatus to the engine house where all members will report for roll-call and cleaning the apparatus.

He shall report at the close of the year to the Assessors, giving a detailed report of each fire and alarm, and make such suggestions as he may deem necessary.

He shall have the general care and charge of the alarm system and see that the same is kept in working order. If the system shall be tampered with he shall report the same to the Assessors with the information he may have, to the end that the guilty parties may be complained of and prosecuted.

First Assistant Foreman. It shall be the duty of the First Assistant Foreman to assist the Chief in the discharge of his duties and report to him any neglect of duty or other matters that are in any way related to the efficiency of the department. It shall be his duty when an alarm is sounded to repair to the locality indicated by the alarm, and report at once to the Chief and assist in getting the apparatus at work with all speed. In case the Chief is not present, the duty of the Chief will devolve upon him, and he will be ready to give the company orders as soon as the apparatus shall each the fire.

The Second Assistant Foreman. When a fire alarm is heard it will be the duty if the Second Assistant Foreman to repair with all speed to the engine house and to take charge of getting apparatus to the scene of the fire, and when the fire is reached to report to and make over to the Chief the apparatus. In absence of his superiors he will succeed to their duties.

Engineers and Assistants. The Engineer shall have charge of the steamer and its management and under the direction of the Chief shall see that it is kept in good condition and ready at all times for immediate service, and if any of the parts are out of order or repair he shall be held responsible, until the same is reported to the officer in charge, or the Chief Engineer. He shall make such repairs as may be possible under the direction of the Chief and shall inform the Chief whenever the steamer may need materials or repairs. He shall instruct the Assistants as to their duties. When the steamer is not in use the Engineer and his Assistants shall be regarded as members of the company.

The Pipemen. It shall be the duty of the Pipemen to lay the pipe (hose) and then direct the streams of water into the fire.

The Hydrant-men. It shall be the duty of the Hydrant-men to attach the hose to the hydrants and turn on the water, as may be needed and to open and close the hydrants as may be directed.

Hook and Ladder Foreman. The Foreman of the Hook and Ladders and his Assistants shall see to it that the apparatus under their charge is conveyed to the fire with the other apparatus. It shall be their duty to care for this apparatus during any fire and to place in position for use as the Chief may direct.

The Lineman will also be a Special Police. His duties will be to care for and protect all pipe lines (hose lines) that may be laid by the company at any fire.

Fire Wardens. There shall be four Fire Wardens appointed by the Assessors. Two of them shall be appointed to perform the same duties as those prescribed for Town Fire Wardens in the laws of the State. It shall be the duties of the other Fire Wardens to have a general oversight of the Corporation with reference to the safety of property from fires. They may be called upon at any time to examine any building or place with reference to danger from fire, and in such cases they shall make a written report to the Assessors. It shall be their duty to make examinations of each residence and other buildings with reference to condition of the fire fixtures, chimneys, etc. Wherever there are defective flues, or other dangerous conditions they shall be reported to the Assessors, and the occupants or owners of the premises shall be notified of the conditions and if within a reasonable time the necessary repairs to make the building safe are not made it shall be the duties of the Fire Wardens to notify the New England Insurance Exchange of the dangerous condition. The regular examinations shall be made as near the first of November as may be.

They shall report at any time on any condition within the Corporation that may be considered unsafe with reference to fires, to the end that prevention is always the better cure.

West Farmington and Suburban Companies.

When there is a general alarm, calling out the West Farmington and Suburban Fire Companies, each company will report for duty on arrival at the fire to the Chief Engineer, and during the fire they will be under his direction.

1998 Harold "Stub" Hemingway with Chief Bob McCleery was recognized for 56 years of service. Stub joined the department in 1942 and served until his passing in 2009, 67 years.

Photo Courtesy of Town of Farmington

Chapter 17

All Appropriations & Charts Courtesy Town of Farmington as published in
Town Annual Reports

FIRE DEPT. APPROPRIATIONS UNDER VICTOR C. HUART TENURE 1939 & 1948

1939

FIRE PUMPER ACCOUNT

Received from other towns	$ 84 00	
Due from other towns	107 00	
Overdrawn	91 21	
		$282 21

Expenses:		
Currier Insurance Agency, insurance	$93 75	
Maxim Motor Company, repairs	1 29	
Forest Fire Payroll No. 1, Porter Hill	8 00	
Currier Insurance Agency, pumper	5 50	
J. W. & W. D. Barker, gas	30 17	
Arthur Locke, repairs	1 60	
Farmington Village Corp., repairs	5 00	
Victor C. Huart, care of pumper	50 00	
Chas. Nickerson, Treas., paid men for fires outside Corporation	87 00	
		$282 21

1948

FIRE DEPARTMENT

Appropriation	$1,000 00	
Receipts	217 25	
Transfer — Overlay	597 07	
		$1,814 32

Expenditures

Warrants drawn	$1,814 32

FIRE FIGHTING EQUIPMENT

Appropriation	$7,000 00

Expenditures

Warrants drawn	$5,202 94	
Unexpended to 1949	1,797 06	
		$7,000 00

AID FARMINGTON FALLS COMPANY

Appropriation	$600 00

Expenditures

Warrants drawn	$600 00

FIRE DEPT. APPROPRIATION UNDER J. BAUER SMALL 1963

FIRE DEPARTMENT APPARATUS RESERVE
Receipts

Balance from 1962	$15,559 08	
Interest on Money	628 58	
		$16,187 66

Expenditures
None

Bal: Deposit in Capital Reserve Account	$16,187 66

FIRE DEPARTMENT
Receipts

Appropriation	$ 6,000 00	
Other Receipts	410 47	
		$ 6,410 47

Expenditures

Payrolls	$ 2,319 25	
Insurance	533 32	
Maint. & Operation of Apparatus	154 14	
Supplies	471 70	
Telephone	345 72	
Rent	965 38	
Maintenance of Radios	232 64	
Repairs	1,030 05	
Postage	39 45	
Transfer	190 01	
Miscellaneous Expense	23 37	
		$ 6,305 03
Unexpended Balance to Summary of Accounts		105 44
		$ 6,410 47

IMPROVEMENT OF FIRE ALARM SYSTEM
Receipts

Appropriation	$ 1,500 00	
Other Receipts	190 01	
		$ 1,690 01

Expenditures

Purchase of Tone Signal System	$ 1,690 01

MISCELLANEOUS PROTECTIONS
Receipts

Appropriation	$ 625 00	
Clinics	131 00	
		$ 756 00

Expenditures

Immunizations	$ 597 47	
Care of Transients	23 00	
		$ 620 47
Unexpended Balance to Summary of Accounts		135 53
		$ 756 00

HYDRANT RENTAL
Receipts

Appropriation	$ 6,720 00

Expenditures

Rental on 40 Rural Hydrants	$ 2,370 00	
Rental on 42 Village Hydrants	2,520 00	
Rental on 17 Suburban		
Water District Hydrants	1,020 00	
Rental on 12 West Farmington Hydrants	720 00	
		$ 6,630 00
Unexpended Balance to Summary of Accounts		90 00
		$ 6,720 00

EXHIBIT D

TOWN OF FARMINGTON
Statement of Departmental Operations
YEAR ENDED DECEMBER 31, 1975

	Appropri- ations	Other Credits	Total Available	Expendi- tures	Balances Lapsed	Balances Carried
GENERAL GOVERNMENT						
Administration	$ 73,535.00	$ 1,648.59	$ 75,183.59	$ 70,791.44	$ 4,392.15	
Community Center Building	15,331.00	5,653.80	20,984.80	16,073.28	4,911.52	
Community Center Gymnasium		3,586.17	3,586.17	2,973.88	612.29	
Town Clerk's Account		19,298.00	19,298.00	13,543.00	5,755.00	
Municipal Building	80,000.00	125,731.16	205,731.16	205,731.16		
PROTECTION						
Police Department	94,247.00	6,474.33	100,721.33	89,430.25	10,591.08	$ 700.00
Fire Department	14,475.00	3,405.29	17,880.29	17,880.51	(.22)	
Hydrant Rental	18,995.00		18,995.00	18,876.25	118.75	
Fire Station Study Committee		6,697.60	6,697.60	5,000.00		1,697.60
Fire Engine		56,019.08	56,019.08	53,250.00	2,769.08	
Fire Department Equipment	5,000.00		5,000.00			5,000.00
Street Lights	24,500.00		24,500.00	24,121.26	378.74	
Civil Defense	841.00	306.00	1,147.00	828.46	318.54	
Workmen's Compensation	6,000.00		6,000.00	4,726.00	1,274.00	
Insurance	4,700.00	112.00	4,812.00	4,583.77	228.23	

EXHIBIT D

TOWN OF FARMINGTON
STATEMENT OF DEPARTMENTAL OPERATIONS
Year Ended December 31, 1976

	Appropria- tions	Other Credits	Total Available	Expendi- tures	Balances Lapsed	Balances Carried
GENERAL GOVERNMENT						
Administration	$ 79,126.00	$ 23,312.40	$ 102,438.40	$ 98,720.91	$ 751.64	$ 2,965.85
Municipal Building	3,056.00	174.42	3,230.42	6,294.74	(3,064.32)	
Town Clerk's Account		22,547.25	22,547.25	15,660.50	6,886.75	
PROTECTION						
Police Department	100,242.00	12,205.90	112,447.90	105,505.66	3,554.24	3,388.00
Fire Department	45,300.00	12,101.33	57,401.33	44,411.50	2,217.83	10,772.00
Fire Station—Municipal Building Construction	110,000.00	305,692.00	415,692.00	415,692.00		
Fire Station Study Committee		1,697.60	1,697.60	1,697.60		
Fire Department Equipment	5,000.00	5,000.00	10,000.00	10,000.00		
Street Lights	25,000.00		25,000.00	24,121.39	878.61	
Insurance and Bonds	9,275.00		9,275.00	8,198.68	1,076.32	
Workmen's Compensation	7,500.00	983.12	8,483.12	8,406.38	76.74	
Civil Emergency Preparedness	886.00	390.00	1,276.00	856.45	419.55	
HEALTH AND SANITATION						
Ambulance Service	17,537.00		17,537.00	16,711.80	825.20	
Sewer Maintenance		113,114.60	113,114.60	113,114.60		
Sewer Construction Reserve		3,330.73	3,330.73	3,330.73		
Sanitary Landfill	12,964.00	6,949.73	19,913.73	24,079.93	(4,166.20)	
Tri-County Mental Health	3,394.00		3,394.00	3,394.00		
Androscoggin Home Health	2,263.00		2,263.00	2,263.00		
Plumbing Permits		786.75	786.75	786.75		

TOWN OF FARMINGTON
STATEMENT OF DEPARTMENTAL OPERATIONS
YEAR ENDED DECEMBER 31, 1977

	Appropri- ations	Other Credits	Total Credits	Debits	...Balances... Lapsed	Carried
GENERAL GOVERNMENT						
Town Manager	$ 30,211.00	$ 336.15	$ 30,547.15	$ 32,191.52	($ 1,644.37)	$
Assessor	25,668.00	3,028.95	28,696.95	30,527.06	(1,830.11)	
Clerk and Treasurer	36,069.00	12,785.90	48,854.90	53,474.04	(4,619.14)	
Community Development		7,500.00	7,500.00	2,661.77		4,838.23
Miscellaneous	36,240.00		36,240.00	35,928.30	311.70	
Municipal Building	144,611.00	49,146.58	193,757.58	161,012.46	32,745.12	
PROTECTION						
Police Department	138,681.00	9,637.33	148,318.33	139,919.12	5,349.21	3,050.00
Fire Department	32,709.00	17,466.56	50,175.56	46,993.99	3,181.57	
Street Lights		30,000.00	30,000.00	29,763.16	236.84	
Civil Emergency Preparedness	1,662.00	288.00	1,950.00	2,264.19	(314.19)	
Hydrant Rental	10,000.00	30,000.00	40,000.00	22,766.60	4,838.23	12,395.17
HEALTH, WELFARE AND SANITATION						
Health and Sanitation	36,288.00	16,171.98	52,459.98	42,162.41	10,297.57	
Ambulance Services	15,600.00	250.00	15,850.00	16,146.00	(296.00)	
Androscoggin Home Health	2,263.00		2,263.00	2,263.00		
Western Older Citizens Council	2,156.00		2,156.00	2,156.00		
Tri-County Mental Health	1,697.00	3,960.00	5,657.00	5,597.00	60.00	
Sewer Maintenance		32,534.18	32,534.18	43,940.35	(11,406.17)	
RECREATION						
Parks and Recreation	86,430.00	15,744.08	102,174.08	103,846.41	(1,672.33)	
Public Library	15,000.00	200.00	15,200.00	15,200.00		
Bureau of Recreation		21,319.63	21,319.63	23,463.21	(2,143.58)	
Swim Program	200.00		200.00	397.46	(197.46)	
Wading Pool	200.00		200.00	142.40	57.60	
Winter Recreation	1,000.00		1,000.00	1,807.32	(807.32)	
Creative Play	800.00		800.00	1,005.57	(205.57)	

FIRE DEPT. APPROPRIATION UNDER ROBERT L. MCCLEERY 1987

TOWN OF FARMINGTON, MAINE
Schedule of Departmental Operations
For the Year Ended December 31, 1987

	Balance 1-1-87	Appropriations	Other Credits	Total Available	Expenditures	Other Debits	Total Expenditures	Balances Lapsed	Balances Carried
GENERAL GOVERNMENT									
Town Manager		58,819.00	4,554.09	63,373.09	62,896.35	237.36	63,133.71	239.38	---
Assessor's Office	2,500.00	33,765.00	1,263.64	37,528.64	33,970.25	2,641.91	36,612.16	916.48	---
Clerk and Treasurer		67,928.00	299.77	68,227.77	65,464.33	356.41	65,820.74	907.03	---
Committees and Events		20,676.00	332.59	21,008.59	17,622.45		17,622.45	3,386.14	---
Municipal Building		25,739.00	299.77	26,038.77	22,122.02	1,389.84	23,511.86	2,526.91	1,500.00
Tax Anticipation Note - Interest		27,000.00	---	27,000.00	24,314.88	---	24,314.88	2,685.12	---
	2,500.00	233,927.00	6,749.86	243,176.86	226,390.28	4,625.52	231,015.80	10,661.06	1,500.00
PROTECTION									
Police Department	818.00	352,502.00	9,836.59	363,156.59	353,085.59	1,042.24	354,127.83	26.76	9,000.00
Fire Department		119,768.00	5,622.29	126,410.29	117,211.68	372.07	117,583.75	4,136.54	4,690.00
Civil Emergency Preparedness		1,083.00	900.00	1,983.00	1,933.88	9.36	1,943.24	39.76	---
Street Lights		51,563.00	---	51,563.00	42,680.09	3,545.85	46,275.94	5,337.06	---
Hydrants		82,000.00	---	82,000.00	81,701.12	---	81,701.17	798.00	---
Insurance		56,213.00	---	56,213.00	38,040.00	---	36,040.00	13,973.00	4,649.16
	818.00	663,129.00	17,358.88	681,723.88	634,652.36	4,969.52	639,621.88	24,362.84	18,339.16
HEALTH, WELFARE AND SANITATION									
Sanitary Landfill		85,680.00	4,439.26	90,319.26	85,803.36	58.43	85,861.79	4,457.47	---
Ambulance Service		24,750.00	70.50	24,750.00	24,750.00		24,750.00		---
General Assistance		3,100.00	18.50	3,170.50	2,086.83	23.61	2,110.44	1,060.06	---
Tree Program		5,000.00	---	5,018.50	4,213.00	82.04	4,213.00	805.50	---
		118,730.00	4,528.26	123,258.26	116,853.19		116,935.23	6,323.03	---
HIGHWAYS									
Parks and Recreation		43,113.00	2,711.02	45,824.02	45,460.76	385.46	45,846.22	(22.20)	---
Community Center		63,544.00	1,709.49	65,253.49	55,734.44	19.99	55,754.43	9,499.06	---
Basketball Court	1,000.00	---	1,000.00	2,000.00		1,000.00	1,000.00		1,000.00
	1,000.00	106,657.00	5,420.51	113,077.51	101,195.20	1,405.45	102,600.65	9,476.86	1,000.00
HIGHWAYS AND BRIDGES									
Town Department	(1,147.80)	518,149.00	50,019.65	594,840.25	548,995.75	32,439.30	581,935.05	(1,147.80)	15,093.00
Public Works Department	26,671.60	518,149.00	50,019.65	593,692.45	548,995.75	37,939.30	581,935.05	(2,162.00)	15,093.00
	25,523.80	518,149.00							
TOWN ASSESSMENTS									
S.A.D. #9		1,129,628.00	---	1,129,628.00	1,129,628.00		1,129,628.00		---
County Tax		116,844.00	---	116,844.00	116,844.00		116,844.00		---
Overlay		52,641.24	2,138.06	52,641.24	---		---		---
		1,299,113.24	2,138.06	1,299,113.24	1,246,472.00		1,246,472.00		---
CLASSIFIED									
Community Development	(2,138.06)	38,552.00	2,138.06	40,343.60	37,552.00	2,791.60	40,343.60		---
Outside Groups	1,791.60	38,552.00	2,138.06	40,343.60	37,552.00	2,791.60	40,343.60		---
	(346.46)								
TOTALS	29,493.34	2,978,277.24	86,215.22	3,093,985.80	2,912,110.78	46,813.43	2,958,924.21	99,129.26	35,932.24

The accompanying summary of significant accounting policies and notes are an integral part of the financial statements.

Edward J. McGuire
Certified Public Accountant

FIRE DEPT. APPROPRIATION UNDER ROBERT L. MCCLEERY 1996 - 1999

SUMMARY OF APPROPRIATIONS, EXPENSES, REQUESTS & RECOMMENDATIONS

Article #		APPROPRIATIONS				1999 Actual Expenditures	RECOMMENDATIONS		
		1996	1997	1998	1999		2000 Requests	Board of Selectmen	Budget Committee
	MUNICIPAL (DEPARTMENTAL) OPERATIONS								
4	**GENERAL ADMINISTRATION**								
	Administration	$150,119	$155,039	$166,285	$168,632	$165,801.23	$179,081	$179,081	$179,081
	Assessor	$72,714	$75,810	$81,427	$85,202	$83,530.91	$87,215	$84,715	$84,715
	Treasurer/Clerk	$133,522	$136,946	$132,932	$147,714	$138,718.64	$148,099	$148,099	$148,099
	Committees & Events	$3,400	$3,500	$12,250	$9,715	$9,166.72	$9,715	$9,715	$9,715
	Municipal Building	$45,361	$47,239	$50,856	$62,276	$61,660.38	$68,644	$60,864	$60,864
	Tax Anticipated Note Interest	$30,000	$22,381	$15,000	$12,000	$7,100.00	$12,000	$12,000	$12,000
	SUBTOTAL - GENERAL ADMINISTRATION	$435,116	$440,915	$458,750	$485,539	$466,074.74	$504,754	$494,474	$494,474
5	CODE ENFORCEMENT	$56,541	$62,572	$76,315	$80,082	$79,096.86	$94,207	$94,207	$94,207
6	POLICE DEPARTMENT	$582,852	$576,436	$616,000	$640,587	$625,069.31	$677,717	$677,717	$677,717
7	FIRE DEPARTMENT	$198,792	$197,136	$196,453	$225,937	$209,873.88	$256,766	$249,627	$229,627
8	**OTHER PROTECTIONS**								
	Street Lights	$69,253	$69,253	$69,945	$73,000	$73,017.53	$73,500	$73,500	$73,500
	Fire Hydrants	$186,531	$186,531	$186,531	$187,211	$187,210.68	$187,211	$187,211	$187,211
	Insurances	$58,145	$55,558	$65,871	$48,657	$42,394.99	$50,500	$50,500	$50,500
	Ambulance	$74,000	$61,043	$15,000	$0	$0.00	$0	$0	$0
	Tree Program	$3,500	$3,500	$4,000	$3,500	$954.00	Transferred	$0	$0
	Traffic Light Maintenance	$0	$0	$0	$6,000	$2,788.62	$9,710	$9,710	$9,710
	SUBTOTAL - OTHER PROTECTIONS	$391,429	$375,885	$341,347	$318,368	$306,365.82	$320,921	$320,921	$320,921
9	RECYCLING	$27,570	$28,003	$29,487	$28,995	$28,270.52	$34,400	$34,133	$34,133
10	GENERAL ASSISTANCE	$15,000	$14,890	$15,000	$10,000	$5,537.20	$10,000	$10,000	$10,000
11	**RECREATION**								
	Parks and Recreation	$104,934	$109,668	$111,085	$114,256	$120,678.74	$127,204	$121,054	$121,054
	Community Center Building	$64,010	$65,196	$57,155	$58,688	$51,820.52	$59,582	$57,082	$57,082
	SUBTOTAL - RECREATION	$168,944	$174,864	$168,240	$172,944	$172,499.26	$186,786	$178,136	$178,136
12	PUBLIC WORKS - OPERATIONS	$600,041	$594,881	$626,058	$658,976	$653,418.67	$682,255	$676,705	$676,705
	SUBTOTAL - MUNICIPAL OPERATIONS	$2,476,285	$2,465,582	$2,527,650	$2,621,428	$2,546,206.26	$2,767,806	$2,735,920	$2,715,920
	$ Increase over preceding year			$62,068	$93,778		$146,378		$114,492

DISTRIBUTION of YOUR 1954 TAX DOLLAR

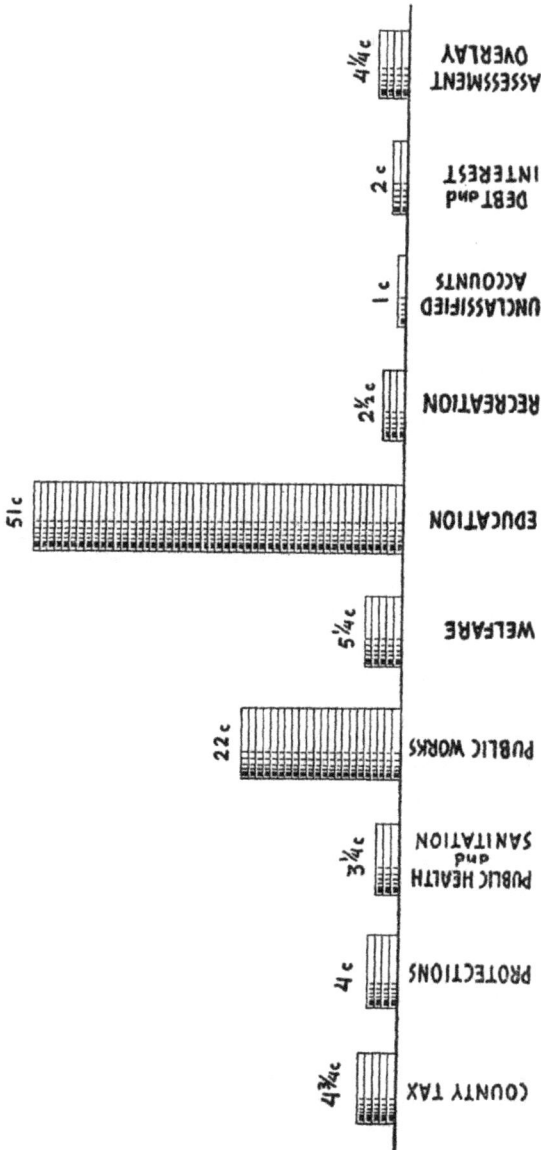

COUNTY TAX — 4¾c

PROTECTIONS — 4c

PUBLIC HEALTH and SANITATION — 3¼c

PUBLIC WORKS — 22c

WELFARE — 5¼c

EDUCATION — 51c

RECREATION — 2½c

UNCLASSIFIED ACCOUNTS — 1c

DEBT and INTEREST — 2c

ASSESSMENT OVERLAY — 4¼c

Costs of Police Department and Traffic Control are not included in this distribution as these functions are financed from parking meter receipts..

Administrative expenses are not included in the above distribution as amounts apportioned for these items were appropriated from surplus from preceding year.

1968 APPROPRIATIONS

%	Category
100.0%	TAX DOLLAR FOR 1968
60.4%	EDUCATION
10.7%	HIGHWAYS
7.7%	PROTECTION — Fire, Police, Hydrants, Street Lights, Civil Defense.
4.6%	UNCLASSIFIED — Social Security, Retirement, Shade Trees, Memorial Day, Tax Discount, Christmas Lighting, Maine Publicity Bureau, Cemeteries.
3.4%	COUNTY TAX
2.8%	GENERAL GOVERNMENT — Officers Salaries, Office Expense, Community Building.
2.0%	HEALTH — Nursing Service, Counseling Service, Sewers, Dump.
2.0%	DEBT & INTEREST
1.8%	OVERLAY
1.8%	RECREATION — Library, Hippach Field, Parks, Wading Pool.
1.5%	RESERVES — Sewer Construction and Fire Equipment.
1.3%	WELFARE — Town Poor, Aid to Dependent Children, Surplus Food.

1973 APPROPRIATIONS

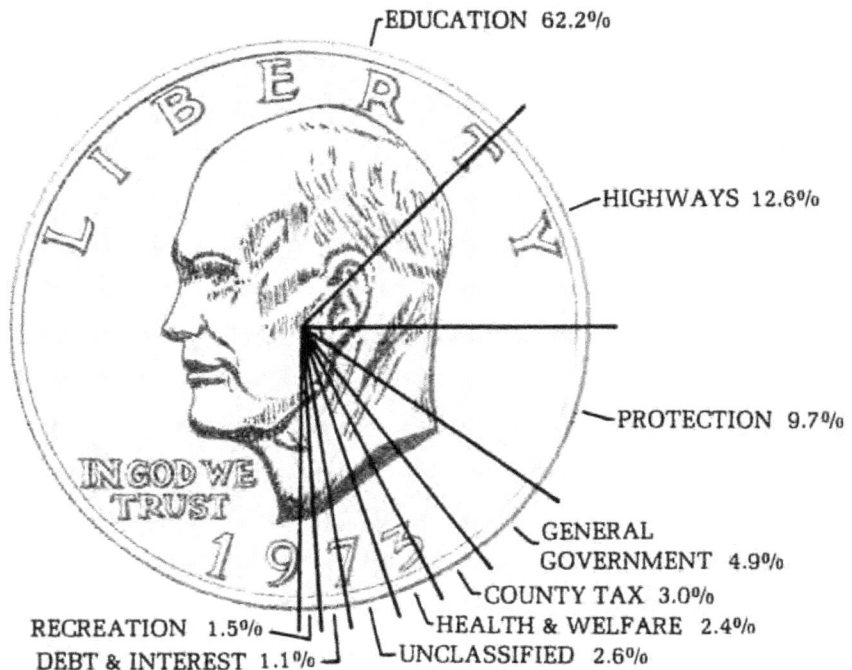

EDUCATION 62.2%

HIGHWAYS 12.6%

PROTECTION 9.7%

GENERAL GOVERNMENT 4.9%

COUNTY TAX 3.0%

RECREATION 1.5%

HEALTH & WELFARE 2.4%

DEBT & INTEREST 1.1%

UNCLASSIFIED 2.6%

PROTECTION — Fire, Police, Hydrants, Street Lights, Civil Defense.

HEALTH & WELFARE — Nursing Service Mental Health, Sewers, Dump, Welfare Assistance, Donated Commodities.

GENERAL GOVERNMENT — Officers Salaries, Office Expense, Equipment, Dues, Attorney's Fees, Property Maps.

UNCLASSIFIED — Social Security, Retirement, Group Insurance, Shade Trees, Memorial Day, Christmas Lighting, Publicity Bureau, Cemeteries, Regional Planning, Abatements, Conservation Commission, Soil Conservation.

RECREATION — Library, Hippach Field, Common, Wading Pool, Swim Program, Creative Play.

WHERE DO YOUR TAXES GO?

MUNICIPAL SERVICES
AND OUTSIDE AGENCIES
$3,075,932
43.93%

SCHOOL
ADMINISTRATIVE
— DISTRICT#9
$2,612,450
51.72%

COUNTY
$258,752
4.35%

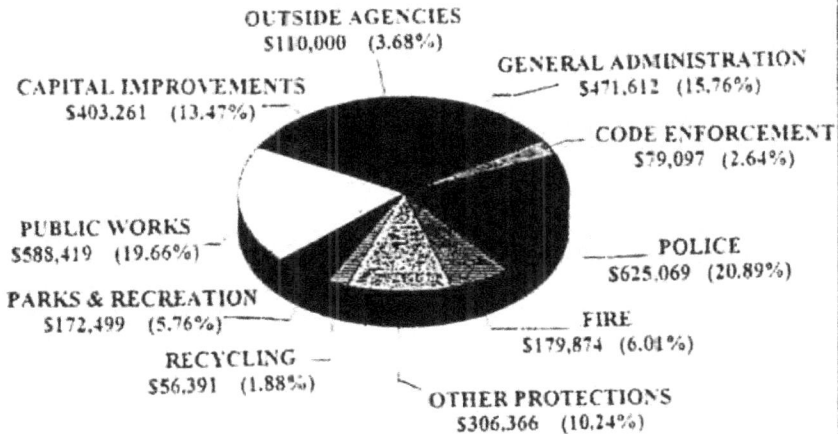

1999 MUNICIPAL
EXPENDITURES

OUTSIDE AGENCIES
$110,000 (3.68%)

GENERAL ADMINISTRATION
$471,612 (15.76%)

CAPITAL IMPROVEMENTS
$403,261 (13.47%)

CODE ENFORCEMENT
$79,097 (2.64%)

PUBLIC WORKS
$588,419 (19.66%)

POLICE
$625,069 (20.89%)

PARKS & RECREATION
$172,499 (5.76%)

FIRE
$179,874 (6.01%)

RECYCLING
$56,391 (1.88%)

OTHER PROTECTIONS
$306,366 (10.24%)

BUDGET COMPARISONS
By Category

■ = 1999 Appropriations   ▦ = 2000 Recommendations

700000

600000

500000

400000

300000

200000

100000

0

General Administration   Code Enforcement   Police   Fire   Other Protections   Recycling   Parks & Recreation   Public Works   Capital Improvement   Outside Agencies

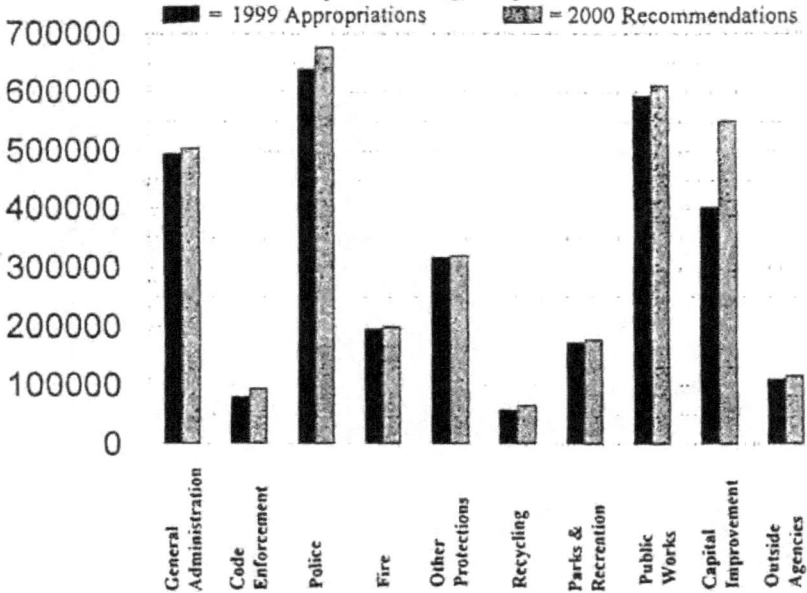

FIRE DEPARTMENT APPROPRIATIONS
1939 - 2000

YEAR

1939	1949	1955	1959	1963	1968	1969	1972	1975	1977
$ 600	1500	2000	2500	6000	8000	8500	10000	14475	32709

YEAR

1981	1987	1989	1993	1995	1996	1997	1998	1999	2000
$ 76465	119788	148075	191469	200883	198792	197136	196452	225937	229627

Chapter 18
Number of Yearly Calls Answered by the Fire Department
1954 - 2000
Courtesy Town Annual Reports

Chief J. Bauer Small

1954 84 Calls

18 calls within Corporation, 4 Suburban Water District, 60 within town limits, 2 New Vineyard.

31 chimney, 17 building, 1 car, 2 brush, 2 grass, 1 short circuit, 5 dump, 1 false alarm.

1955 44 Calls

8 inside Corporation, 2 Suburban Water District, 31 within town limits, 1 New Vineyard, 1 Chesterville. 18 chimney, 1 car, 5 building, 1 oil burner, 5 dump, 1 brush.

1956 60 Fire Calls - Mostly minor fires with 23 being chimney fires.

1957 48 Calls

Town: 11 chimney, 4 buildings, 2 oil burners, 1 spilled gas, 1 tree fire, 1 tie culvert, 1 milk tank, 1 sawdust pile, 9 grass, 2 dump, 3 woods, 2 auto Suburban Water District: 2 chimney, 2 grass, 1 smoke scare.

New Vineyard: 1 chimney, 1 building; New Sharon: 1 building; 1 false; Chesterville: 1 building.

1958 57 Fires

11 in Corporation, 2 in Town of Industry, 1 Chesterville.

7 building, 17 chimney, 4 grass, 5 auto, 2 false, 4 oil burner, 2 dump, 1 air blower.

1959 28 Calls - All outside the Corporation

11 chimney, 2 faulty wiring, 4 building, 2 woods, 2 hot fat burning on stove, 1 dump, 1 flooded oil burner, 1 false, 1 grass, 1 smoldering cigarette, 1 call to Chesterville, 1 call to New Sharon.

1961 73 Calls.

Chimneys and flooded burners headed the list, but had 3 serious fires, these 3 nearly involved fatalities.

1963 62 Calls

Chimney fires predominate. One serious fire at Hubert Knowlton on Perham Street.

1966 59 Calls

Most serious: Loss of barn at Voter Farm and Farmington Supply Company along with chimney and oil burners.

1968 69 Calls

Chimney and grass fires tied with 13 each. Others were usual - cars, oil burners, etc.

Chief Forest Allen

1972 69 Calls. 5 Out of Town

1973 79 Calls. 69 Local, 10 Out of Town

1975 73 Calls

10 car and truck, 13 chimney, 7 house, 5 business, 5 grass, 3 dump, 4 gas spills, 5 false, 4 UMF, 2 flooded burners, 4 car accidents, 2 power lines down, 1 television set, 6 outside calls - 4 to Industry, 2 to Chesterville; one major fire resulting in destruction of Drummond Hall Block.

1976 74 Calls

9 car and truck, 3 business, 4 accidents, 16 chimney, 3 UMF, 9 house (2 fatalities), 6 false, 2 flooded oil burners, 4 grass, 2 woods, 1 call to FCM Hospital, 2 barns and sheds, 6 outside fires, 7 miscellaneous - 1 to Industry, 1 to New Sharon, 1 to Wilton, 2 to Temple, 1 to Chesterville.

Chief Robert McCleery

1977 135 Calls

Out of town calls to Chesterville, Industry, New Vineyard, Weld, accidents, cars, false, grass and woods, oil burners, commercial (Maine Dowel & Macomber Mill), house and miscellaneous calls for overheated stoves, electrical outlets, electric motors, downed power lines, transformers, sawdust piles, wood piles, and paper dispenser at Mt. Blue HS.

1978 160 Calls

Out of town calls, highway vehicle accidents (cars, trucks, cycles), grass and woods, houses, overheated stoves, electrical outlets, warehouses, miscellaneous, and training. Chimney fires continue to be most calls due to improper wood burning and sooty chimneys with an increasing number of calls to inspect chimneys, building, and homes.

1979 103 Calls

Chimney fires largest number of calls, 4 homes (2 mobile) were destroyed, several others serious damage, service calls, false, accidents, vehicles, oil burners, grass and brush, mutual aid, and aircraft crash.

1980 Record Number of Calls

65 chimney fires contributing the greatest number.

1981 Record Year for Calls.

Chimney fires receiving the largest percentage (47%) followed by service calls, vehicle, structure, grass, mutual aid, false alarms, and ambulance assistance.

1982 Calls dropped by 30 from previous year

Chimney fires are still receiving the largest percentage, followed by smoke scares, structures, cars, false alarms, service calls, grass fires, and mutual aid. Dollar value of property loss was the highest in several years. Two structural fires resulted in loss of life of 2 adults, a teenager and young child.

1983 168 Calls

Very high fire loss in buildings and stock this year, totaling nearly a million dollars. 13 structure fires in Farmington with United Lumber (Starbird Home Care Center) the largest. Several of these required mutual aid from Temple on stand-by at Central Station, air bottles and manpower from Wilton with Chesterville and New Sharon covering the Falls area. Farmington FD responded to 12 mutual aid calls to Chesterville, Industry, New Sharon, New Vineyard, Strong, Temple and Wilton. Wilton and Temple responded to aid Farmington FD 6 times.

45 chimney, 21 vehicle, 10 grass, 13 false, 9 smoke scares, 7 accidents, 1 assist Keegan Ambulance personnel, 1 extrication, 1 wood fire, 30 miscellaneous calls of electrical malfunctions, trees on power lines, stove and furnace blow back, gasoline and chemical leaks and vapors, vandalism, personal assistance, debris, cellar pumping, and others.

1984 More calls than any year since Chief McCleery became chief

11 structure fires resulting in $150,000 in loss with some requiring mutual aid from Wilton, Chesterville and Temple, 8 extrication, 54 chimney, 11 vehicle, 6 grass and brush, 15 false alarms and system malfunctions, and numerous miscellaneous and service calls; trees on power lines, gasoline and chemical leaks or spills, debris, cellar pumping, and others.

1985 Busy year, more calls for various incidences

11 structure fires resulting in $255,500 property loss, 45 chimney, 12 vehicle, 9 accidents and or extrication, 9 grass and brush, 32 debris, trash, smoke, furnace, stove, CMP line problems, electrical and gasoline spills, 29 false alarms, 13 miscellaneous and service.

1986 200 Calls

Structure fire at Tom Williams Farm RT 27 in January, Larry Tracy home on Seamon Rd in February, and 3 severe structure fires in May - Luce house on Lower Main St., Sweetser's Mill RT 43 and Kolreg mobile home on back Temple Rd. In May, Robert Hunter home by green bridge in Farmington Falls was destroyed. These fires resulted in over $200,000 of property damage.

41 chimney, 8 vehicle, 12 accidents and or extrication, 8 grass and wood, 16 debris, dump, tires, 15 false alarms or alarm malfunctions, 14 smoke, stove, and furnace problems, 6 electrical related, 10 spills and vapors, 13 cellar pumping, 15 service calls.

During the flooding in January, the Fire Department with boats rescued 4 people stranded at the Diner and 7-11. Fire Department received mutual aid several times as well as responded to neighboring towns several times. The most serious and destructive fire out of town was the Davis Farm in September.

1987 Record Number of Calls

15 structure fires resulting in approximately $195,000 of property loss.

34 chimney, 16 vehicle, 16 accidents and or extrication, 5 grass and brush, 11 dump and debris, 17 false alarms and system malfunctions, 15 smoke and furnace problems, 6 electrical related, 9 gasoline spills, 13 mutual aid, 71 service or miscellaneous, mutual aid calls.

The Fire Department was involved in the April 1st flood of 1987. This included rescue of people by boat from businesses and homes on Intervale Rd. An attempted rescue of a person at Farmington Falls resulted in a near tragedy for 2 of our firemen at Farmington Falls. 36 cellars and treatment plants had to be pumped out.

1988 Down 80 Calls from 1987

34 chimney fires. More citizens have gone back to oil or some other type of heat and those that are burning wood have better heating appliances, new chimneys, and are doing a better job with dry wood.

10 serious structure fires resulted in property loss exceeding $181,000. Most serious problem this year was 34 false alarms or system malfunctions. Most emergency calls were well below the previous year. Other calls, service or miscellaneous, ran about the same.

1989 Relatively Quiet Year

Chimney fires are down, citizens are burning less wood and using other types of fuel for heat.

10 structure fires with property loss exceeding $157,500 (3 totaling more than $30,000) 32 chimney, 10 vehicle, 11 accidents and or extrication, 20 grass, brush, and dump, 26 false alarms and system malfunctions, 19 smoke, stove, and electrical problems, 47 mutual aid, spills, service and miscellaneous.

Farmington also received mutual aid from Chesterville, New Sharon, Industry, Temple and Wilton.

1990 209 Calls

194 emergency, 15 structure with property loss of $305,000

Farmington Fire Department and the surrounding towns have an excellent mutual aid response between them and the cooperation is excellent.

In August, the Farmington Fire Department was involved in its first Hazardous Materials spill involving a tank truck loaded with 6000 gallons of Black Liquor. Our mutual aid people responded to assist along with International Paper Co., DEP, and RST Hazardous Material Team.

1991 232 Calls

189 emergency, 20 service, 23 training

1993 271 Calls

234 emergency, 37 service: 16 structure, 27 chimney, 13 vehicle, 49 accident and or extrication, 4 grass, brush, and debris, 33 false alarms or system malfunctions, 36 smoke, stove, and electrical related fires, 9 gasoline and oil spills, 5 miscellaneous.

Fire Department responded to 24 mutual aid calls and received mutual aid 19 times.

1994 290 Calls

231 emergency, 34 service, 25 training

18 structure, 23 chimney, 20 vehicle, 44 accident and or extrication, 12 grass, brush, and debris, 25 false alarms or system malfunctions, 26 smoke, stove, and electrical related, 15 gasoline and oil spills, 24 miscellaneous, and several mutual aid.

1995 282 Calls

240 emergency, 20 service, 22 training, 15 structure, 13 chimney, 10 vehicle, 75 accidents and or extrication, 10 grass, brush, and debris, 22 false alarms or system malfunctions, 42 smoke, stove, and electrical related, 14 gasoline and oil spills, 9 miscellaneous, and several mutual aid.

1996 320 Calls

272 emergency, 20 service, 9 structure, 13 chimney, 11 vehicle, 95 accidents and or extrication, 15 grass, brush and debris, 17 false alarms or system malfunctions, 24 smoke, stove, furnace, and electrical related, 29 gasoline, oil and hazardous spills, 25 miscellaneous, and mutual aid was given and received.

1997 252 Calls

208 emergency, 27 service, 16 structure, 12 chimney, 12 vehicle, 65 accidents and or extrication, 8 grass, brush and debris, 19 false alarms or system malfunctions, 27 smoke, stove, furnace, and electrical related, 14 gasoline, oil, and hazardous spills, 35 miscellaneous, and several mutual aid.

1998 288 Calls

January was an eventful month with a total of 53 calls.

1998 started off with the Ice Storm. Power was out for days and weeks in some places; trees and power lines were down; Firemen assisted with removal of trees from roads and power lines and helped assist needy citizens to the Community Center for shelter. Some firemen stayed at the shelter and helped with food preparation. Firemen also hauled water with an engine to farms to water livestock.

June was another busy month due to heavy rain and flooding. Extensive damage was done to roads, fields and crop land. Fireman patrolled flooded roads and streets in and around Farmington.

Farmington Rescue was called twice by the Warden's Service to assist with the rescue of injured climbers on Tumbledown Mountain in Weld.

260 emergency calls which included structure fires, vehicle fires, accidents and extrication, grass, brush, and debris fires, false alarms or alarm malfunctions,

smoke, stove, furnace, and electrical related calls, gasoline, oil, and hazardous spills, mutual aid requests from other Fire Departments, and search and rescue.

1999 274 Calls

237 emergency calls which included structure fires, vehicle fires, accidents and extrications calls, hazardous spills, false alarms, mutual aid calls, furnace calls, debris fires, electrical related fires and rescue calls. There were 2 structure fires resulting in total loss: The Brackett House in West Farmington in February, and the Abbott House in June, previous location of MBNA.

2000 255 Calls

250 emergency, 5 service

150 Calls First 6 Months: 9 structure, 4 chimney, 8 vehicle, 41 accidents, 2 extrication, 1 grass, 3 smoke, 5 carbon monoxide, 2 hazardous spills, 3 false alarms, 64 miscellaneous, 7 mutual aid, and 1 training.

Chief Terry S. Bell, Sr.

105 Calls Second 6 Months: Structure, vehicle, accident, extrication, alarm malfunctions, Haz-Mat, gasoline & oil spills.

From 2000 - 2016 Fire Department Calls have increased significantly each year.

2015 & 2016 430+ Calls - Fire Department Record # of Calls.
(per Deputy Chief S. Clyde Ross)

NUMBER OF CALLS PER YEAR 1954 - 2016

YEAR

1954	1955	1956	1957	1958	1959	1961	1963	1966	1968	1972
#84	44	60	48	57	28	73	62	59	69	69

YEAR

1973	1975	1976	1977	1978	1979	1980	1983	1986	1990	1991
#79	73	74	135	160	103	65	168	200	209	232

YEAR

1993	1994	1995	1996	1997	1998	1999	2000	2015	2016
#221	290	282	320	252	288	274	255	434	430

Chapter 19
FIRE DEPARTMENT HISTORY UPDATED
By Deputy Chief S. Clyde Ross
August 26, 2017

2000

Fire Chief Robert L. McCleery retires after serving in that position since 1977; He was replaced by Deputy Fire Chief Terry S. Bell, Sr. Chief Bell assumes his new position as the first full time Fire Chief and will have personal truck for use during his tenure.

2001

The "Dream" comes true. Chief Bell and Director of the Mt. Blue High School Foster Technical Center Anne Deraspe put together plans for the first Fire Fighting Class for High School Students. Jack Berry was the instructor with assistance from a variety of State of Maine Fire Instructors. There were 16 students and adults in this class. September 11th was a devastating day for these students as well as all of us when the New York City Trade Centers collapsed after being struck by aircrafts; this was a terrorist attack on the United States.

2002

Former Fire Chief Robert L. McCleery passed away in July with his funeral services held at the Franklin County Agricultural Society's Starbird Building. The service was attended by fire fighters from a large number of departments in Maine.

Purchase of Engine #1 ($371,00.00) from Pierce in Wisconsin. This has a 6000 watt Light Tower on it, dump for Speedi-Dri compartment, overhead doors, with "vault" compartments on top for storage of water rescue items, ropes, hand tools etc. The Hurst Extrication Tools are all mounted on this unit, 3 hoses for tools are a fine addition.

2004

Deputy Fire Chief S. Clyde Ross was elected president of the Franklin County Firemen's Association at the annual meeting held here in Farmington.

2005

Purchase of a Bauer Air Filling System with Grant funds ($32,000.00) from Grant funds, this replaced the Poseidon unit that we had had for several years. Air filling station with DOT Air Bottles was also purchased for the Ladder Truck. Unit was placed in the Generator Room and the Storage Room.

2001 Farmington Fire Rescue in NYC To Assist with Recovery Efforts after 911
Top: Chief Terry Bell with NYC fireman
Bottom: Melvin Bard & Richard Knight 2 of 15 Farmington Firemen who made trip to Ground Zero

Maine Dowel Fire January 29, 2002
Photos Courtesy Don DeRoche Photographer Pat Durham
Chief Terry Bell and Deputy Chief Clyde Ross in middle picture.

Mt. Blue senior is Farmington's newest firefighter

By BETTY JESPERSEN
Staff Writer

FARMINGTON — The Farmington Fire Department's newest member is a young woman who was a student in the first class of the Firefighter 1 course now being offered regularly at the Foster Applied Technology Center.

Selectmen this week approved Chief Terry Bell's recommendation to hire Mt. Blue High School senior Abby Davenport, 18, of West Farmington. She has been volunteering at the station since last summer and was officially sworn in Wednesday.

"Abby is energetic and enthusiastic and will be a great asset. She knows her physical limitations but there is no doubt she can pull her own weight," said Deputy Chief Tim Hardy. "And the big plus is that she is going to the University of Maine at Farmington so she will be with us for four more years."

The Firefighter 1 course at FATC, the first in the state, is taught and co-sponsored by area fire departments. Each student is assigned a mentor from their hometown department. But according to the program's organizers, all the departments have taken their fledgling firefighters-in-training under their collective wings.

"I never thought about being a firefighter until last summer when I decided to sign up for the course and then I really wanted to get into the department," Davenport said.

"But I didn't think I had a chance because I thought you had to be in

> "They told me they would teach me how to use the chain saw but we haven't gotten that far yet."
>
> Abby Davenport,
> Farmington firefighter

the 'loop' and I had no family member in the department and I was a girl."

But that all changed when she jumped into the program and saw how the students were encouraged to stay with the program and become firefighters. The course was challenging, and FATC students must complete the same requirements as adult firefighters but in half the time.

Davenport, the daughter of Gordon and Kathy Davenport, said she is often at the fire department after school, waiting to respond to a fire or accident and in the meantime, helps manage the computer system and has already set up the department's Web page.

"They told me they would teach me how to use the chain saw but we haven't gotten that far yet," she said with a smile.

The Farmington department re-

Please see HIRED, B3

• Hired

Continued from B1

quires firefighter candidates pass an oral interview and an agility test as well as have Firefighter 1 certification. Davenport discovered the most daunting challenge was the requirement she jump 4½ feet wearing full turnout gear and carrying an air pack — adding a total of 50 extra pounds to her slight frame.

"I practiced jumping without my gear, then with it, and developed a technique and went over it repeatedly until I got it," she said.

The thrill of being accepted as a firefighter is tremendous, she said. "Everyone accepts me here, and I am so comfortable with all of them. It is like a big family."

The distinction of being the first female firefighter goes to Sheila Landry, who was with the department years ago, Hardy said.

Davenport will be on probation for one year. She is also taking the new emergency medical technician course at FATC, which she will complete in June.

Betty Jespersen — 778-6991
jespersen@prexar.com

Staff photo / BETTY JESPERSEN

Abby Davenport, 18, of West Farmington, is the newest member of the Farmington Fire Department. She was a student in the first class of the Firefighter 1 course now being offered regularly at the Foster Applied Technology Center.

Article Courtesy Don DeRoche
and
Franklin Journal March 25, 2002 Edition
Staff Writer Betty Jespersen
Pictured: Abby Davenport

2006

With all the new Mandated Training items, the first Saturday in January is designated as Training Day; lunch is also serviced during this day's training classes. Classes were held at Mt. Blue High School with local doctors and nurses doing the Medical Assessments.

2007

Purchase to Tower #3 ($812,000.00) again from Pierce; tower unit with a piped line to the bucket, numerous lights, overhead doors for compartment, airway to bucket, numerous storage compartments, air conditioned this time, nice lighting system with adequate generator and accessories. Ladder 1 was sold to Livermore Falls Fire Dept.

Per Diem Fire Fighters were hired after March town meeting and have worked out very nicely. They receive only an hourly wage with no benefits. New Fitness Training Equipment purchased with Grant Funds.

William Goldfeder came in October to speak to some 85 fire fighters at the Mt. Blue High School.

2008

Purchase of three new overhead doors to replace the 1980's "fuel saver doors" that no one ever liked and we never knew what was saved, if anything for fuel. They look very nice. The Squad Truck was outfitted with a "skid tank", hose reel, foam unit and various hand tools. Generator and pump are still on this unit. Built a new Storage Shed in back of the Station to house numerous pieces of surplus equipment.

New floor put in the Grandstand Fair Booth. No Easter Egg Hunt this year, too many problems.

Two training classes provided by outside Instructors, Fire Fighter Survival Techniques and Rural Water Shuttles (Farmington and Strong) classes open to the County membership and well attended. Fire Fighter Eugene Mosher died in July and his services were appreciated for years of service. Engine #6 Mack has been sold.

2009

Purchase of the Rescue #1 Vehicle, $7,500.00, (1997 International unit from Westbrook); this is equipped with the Air Filler System (4 tanks), hand tools, spare air bottles, nice heater system and "office" space if needed. Two more overhead doors were installed (1 front and 1 rear of station) to complete the new door process. The All Hazards Pre-Plan Computer Program was purchased and units installed in Engines 1-2-Tower #3, Chief's truck, EMA Truck and Strong's Chief's Truck. These have a variety of programs installed (pre-plans, resource lists, hydrants locations, Knox Box locations, and more to be added). This is Grant funds being put to good use.

We were saddened in the summer and fall by the passing of Senior Captains Harold "Stub" Hemingway and Senior Captain John "Jack" Bell. They have been dedicated Fire Fighters for many many years and their contributions will be missed.

The "big one" this year was at the Hazel Thompson senior citizen complex in August when fire broke out in a first-floor apartment. Forty occupants were removed and placed in alternative housing that night. There were no injuries and people responded very well to the situation.

A new Ford Pickup truck, 4X4 Crew Cab, was purchased locally for Chief Bell and fire department business at a very good deal.

NIMS training for officers continued this year in different locations.

Stephen Bunker was elected President of the Franklin County Firemen's Association. We continued working on radios, the repeater system and Narrow Banding and Licenses.

2010

We started off this year with a big one; the new Medical Arts Building at Franklin Memorial Hospital caught on fire January 30th. It was an early cold morning and was reported as "water leaking through the ceiling". It was and there was fire in the attic area. Several mutual departments assisted in the extinguishment of this $9 Million loss, estimated.

Traditions Training came again this year for the two days of intense and very educational training classes participated in by several of our mutual aid departments. The old Keith House, corner of Perkins and Quebec Streets, was the scene for this training.

A first for the department this year, we didn't open any Food Booths at the annual Franklin County Fair. It was voted by the members to fore-go any fair project this year. Phil Allen and Richard Chabot did operate one of the booths and made a donation to the Benevolent Association.

The summer outing was again enjoyed by family members with all the traditional food items being served. For the first time in many years no one attended the New England Fire Chiefs Convention in Springfield, MA. Chief Bell got a new pickup truck this year; the Recreation Dept. purchased the old one.

John O. "Jack" Bell   11/23/28 – 8/8/09

Jack was born in Brooklyn, New York and, when he was 14, his family moved to Farmington where he graduated from Farmington High School. He worked for Forster Manufacturing and later Hannaford Oil Company. In 1981, Jack obtained his Master Plumber's License and started Jack Bell Plumbing. He was a dedicated member of the Farmington Fire Rescue Department for 49 years. Jack served as a delegate of the Maine Federation of Firefighters, and his company supplied bikes to the winners of the annual Easter Egg Hunt for the Benevolent Association. When the Fire Department was located on High Street, he helped organize the Christmas Party, and the annual dance at the Community Center, which raised funds for the Department. Jack served on the Truck Committee, and he supported the new equipment methodology for safety and training. He mentored the younger members of the Department, and was well respected by all as he led by example and with encouragement. His sons Terry and Mike are following in his footsteps.

Harold H. "Stub" Hemingway   09/21/15 – 09/15/09

Stub was born in Farmington and graduated from Farmington High School in 1934. He served in the U.S. Army from 1945 to 1946 in Saipan. Stub worked in public safety as a Franklin County Deputy Sheriff, security officer for Franklin County Superior Court, reserve officer for the Farmington Police Department, and for the University of Maine Campus Police. He also worked for Coca Cola for about 28 years. Stub was a member of Farmington Fire Rescue Department for 67 years, and was instrumental in the move from the old High Street firehouse to its present location. His knowledge and experience in public safety and security was an important asset to the Department. Being a senior firefighter, Stub mentored the younger members, and encouraged them to continue their safety equipment training. Stub enjoyed gardening and shared his produce graciously. He provided homemade pickles for all the Department events. Stub enjoyed the outdoors, and was an avid cribbage player and Red Sox fan.

IN MEMORIAM
Courtesy Farmington Fire Department Town of Farmington

2011

Engine #2 was sent to Portland for repairs after a continuing braking problem was reported. In February, the department was given the OK to start a Junior Fire Fighter Program; we had 4 young men come into the program once it was established. The SHAPE AWARD inspection was done in March and once again, with a few modifications, we received a fine review results.

December 29th saw the Knowlton Corner Horse Farm burn to the ground in the early morning hours, no animals were lost and the home was saved by a very direct attack with limited water supply. Several Mutual Aid Companies responded with tankers, apparatus and personnel. There were no injuries and 3000 feet of LDH was laid form the new Reservoir near the High School. It was a cold and windy morning, 0345.

2012

We have started off the year with a very busy call volume, numerous local and Mutual Aid calls so far. In February were once again called to Strong for a fire in the Pellet Mill, a lot of smoke and water damage. Early spring in March/April saw several grass/brush fires but no property lost. On April 14th, early morning again, we were call to the Saw Mill Fire on Donald Smiley's Knowlton Corner Road property. Again, Mutual Aid was a big help, tanker shuttle necessary. On April 27th we took part in the Mock 10-55 on South St. sponsored by the UMF and NorthStar Ambulance Service. Philip R. Allen resigned after 33 years of Dedicated Service to the Farmington Fire Rescue Dept. A new Hurst "Cutter/Spreader" has been purchased to help with the newer car metal components.

Philip Allen retired during the summer after 33 years of service to the town. The department operated two Fair Booths this year and made a fine profit. The Federation Convention has been award to Farmington for 2014, this is our third opportunity to host this event. A new Squad Truck has been delivered and we are awaiting the body arrival.

2013

This year saw an unusual number of fire calls for assistance from the public. We were kept busy with weather conditions, rain storms, winds and snow/ice situations. It was also interesting that the new Squad Truck body arrived and through the efficient work of a number of the members we were able to get the unit completed and operating in a manner that was found to be very helpful. The department continues to participate in numerous community events with the BBQ, the Fair and eventually attending the State Fire Fighter Convention where we were awarded the site for 2014.

In July the department responded to Lac Megantic, Quebec for a train derailment. 73 railroad cars carrying crude oil caught fire and the explosion and following fire killed 47 citizens of that community. Later in October a fund-raising event was held at the

University of Maine in Farmington for the Lac Megantic sister city, several citations and recognitions were also presented. The event was well attended by the community and surrounding areas.

Phil Allen and Greg Roux retired from the department after 33 and 19 years respectfully; they have contributed much to the town over the years.

The department has realized that the active membership numbers have decreased for a variety of reasons and how we can attract new members becomes a serious problem. The Selectmen have been informed and more attention is now being considered in the situation.

2014

Resignations still continue, Capt. Richard Knight and Richard Chabot have retired after 35 and 25 years respectfully to attend other commitments. Winter has seen us very busy assisting the highway department and taking care of downed trees and water conditions. Snow and ice build ups have been unusual this year. We are continuing to address the shortage of fire fighters available for call. The convention committee is preparing for the upcoming State Convention in September.

At the recent Town Meeting, it was voted to allow the department to have three Per Diem fire fighters on duty Monday thru Friday and two during weekend days, this will give the town better coverage.

During January, there were several storms and some serious water issues that the department was called upon to assist with traffic, placing traffic cones and helping to free drainage basins of ice and snow. We had 62 calls this month. Winter continued to bring us a "true Maine" weather display, rain, sleet and more snow and ice. Spring came in a hurry once conditions changed and there was no flooding this year.

In July, we had two structure fire caused by lightning strikes. No one could remember when this had happened in the past. Mutual Aid departments were helpful in supplying need manpower and apparatus at these fires. High temperatures and humidity were also factors that made these fires difficult to fight.

Work has been taking place on the Convention, selling Ads, seeing different businesses/organizations, ordering needed supplies and making final arrangements for the numerous events that will be taking place.

The first weekend in September saw the Maine Federation of Fire Fighters Convention here in Farmington, the third time we had hosted since 1985. The weather was again excellent and the turn out for the three days was outstanding. We sold out on the items for sale and the banquet and dance drew larger than expected crowds. In all the weekend was a huge success. Thanks to all who participated in making the arrangements and those attending.

Fair Week was well attended also and our food booths did another outstanding job serving food and assisting the fair goers.

Engine I has had its share of mechanical work done on it this fall; the generator for one and other items have been worked on and brought back to good operating function.

2015

Winter started off with plenty of almost bare ground after the early snows in November and December that caused numerous accidents and hardship for people. Icy conditions and snow were hard to deal with. Late January and early February saw more snow that most of us really wanted. Accidents were the major calls during this time. Temperatures so far this winter have been on the very cold side.

Former Deputy Chief Glenwood Farmer passed away June 13th and the department was saddened by this loss. He was a mentor to many of us and his sense of humor will be remembered.

Engines have been tested as well as the Tower, all passed their tests with good results. The Benevolent Association put on a BBQ on July 3rd at the Narrow Gauge Cinema area and used the "reservation" idea this time. It worked very well and the chicken was all sold. Once again, the annual Fair Booths were in operation and a successful year for each of them. Funds from these are used for Scholarships and other items for the Department (a new kitchen stove, items for the Side by Side and good food now and then).

In October, we had two structure fires (Bates and Jalbert) within a span of 6 hours. Both were total losses as they were out of the hydrant district. Mutual Aid was again very much appreciated and helped with water shuttle and manpower. The John Deere Side by Side, purchased with Homeland Security Funds, received additional equipment with the assistance of the Police Department and Fire Department. This is a "going" unit, has a trailer to transport it where needed and had been used in Mutual Training exercises. The department has been busy as usual and again had a record number of calls, 434 in total.

2016

As we start of the new year, again we are having a numerous calls even though the weather has been very uneventful so far in January. The big item so far has been the Town Budget presentation. Chief Bell has proposed that we hire four full time fighters working 12 hours shifts with one per diem Monday thru Friday. This has generated a great deal of discussion in the administration and with the selectmen. We currently have 25 members and the average age is 53 years old. There have not been many people submitting applications for membership in the past months and crunch time is on us so to speak. At the Annual Town Meeting March 28th the town voted to hire four full time fire fighters. Applications have already been arriving and there are two possible hires from within the department as of April 5th.

On May 1st three full time fire fighters were hired; Captain Timothy D. Hardy,

Scott Baxter, and Joseph Hastings. May 3rd Michael Melville and Scott Baxter were promoted to the rank of Captain by Chief Bell.

Seth Abbott was also hired as a full-time fire fighter; he resigned in December after a short employment period.

The Dept. answered nearly 430 calls again this year. The Dept. continues to hold Mutual Aid Training with nearby towns and is still working on the "basics" of fire-fighting as the primary objective. Prevention activities are a constant with us as we work closely with the schools, senior centers, the University and numerous businesses ask for periodic training sessions.

2017

The year has started off with good snow and ice conditions making travel in the area a challenge at times. Budget preparation this year will include the 4 full time positions for 52 weeks compared to 39 last years, this will increase the need for more funding. We have assisted other towns with Mutual Aid for structure fires in January and early February.

Recently a new "apron" was installed in front of the Fire Station, hopefully this one of concrete will last longer than the previous ones. For the most part things have been quiet during the summer and all. The Chiefs did attend the Bethel Maine Fire Chiefs Meeting and seminars during that time. We always have a good time and can learn from the fine speakers. Engine 1 has been out of service for a short time but is now back in service.

FFD Photos Courtesy Town of Farmington

Deputy Chief Clyde Ross, Chief Terry Bell
Pictures Courtesy Town of Farmington

Phil Allen

Richard Knight

In Memoriam

Dr. Paul A. Brinkman
1932 – 2002

Dr. Brinkman's community involvement included serving on the Board of Trustees of the Pierce House, the UMF Gold Leaf Program, and the volunteer program at SAD 9, to name only a few. He received numerous awards for community service such as the Community Health Leadership Award in 1997. He offered his professional assistance to Town committees, to the Police Department, and to the Fire Rescue Department. He will be remembered for his unselfish commitment to civic responsibility.

Dr. Phillip B. Chase
1916 - 2002

Dr. Chase served as family practitioner in the Farmington area off and on from 1946 until his retirement in May of 1995. He was an active outdoorsman, a great historian, and a founding member of St. Luke's Episcopal Church. He sought the finer qualities in individuals, and his knowledge and compassion were felt around the globe. He was a devoted family man who always maintained a positive attitude. We remember his generous professional assistance in committee work and in service to the Town.

Robert L. McCleery
1922 – 2002

Bob was noted for his leadership abilities and was recognized with many honors for outstanding community service. He was Chief of the Farmington Fire Rescue Department from 1977 until his retirement in 2000 and was Maine Fire Chief of the Year in 1998. His philosophy of life included maintaining "stewardship" of the land, love and support of family, and strong community involvement. Bob was a leader, a mentor, and a lover of all that life provided.

In Memoriam Courtesy Town of Farmington 2002 Annual Report

Chapter 20

The Life of Robert L. McCleery
1922 - 2002
A Daughter's Personal Reflections

Dad passed away in the early morning hours of July 22, 2002. He had been active and doing well until a week prior when he was taken ill at a Sunday afternoon family gathering at Clearwater Lake. Dad was diagnosed with peritoneal carcinomatosis caused by mesothelioma which turned out to be a very fast-moving cancer, and he was gone within 5 days. We were not prepared for this. He was in rare form at his 80th birthday party the month before; had baled hay a couple weeks prior; had two new knees; his heart was good; and he was doing great. We all thought he had another good five to ten years.

Dad was our rock, and his passing has left a huge hole in our lives as has the passing of our mother in 2011. Mom was our advocate. If there was a problem to be addressed, she would be there to address it. She was a social worker at heart, always ready to help anyone in need, especially a child. Dad was my hero; he was always loving, supportive and encouraging. If I doubted if I could do something, it was Dad who would say: "How do you know? You haven't tried."

In 2002, God must have had need for a few good men as a month prior to Dad's passing Cape Elizabeth Deputy Fire Chief L. P. "Jimmy" Murray died, and several months later John Harmon, Sr., Scarborough Fire Chief 1982 - 1984, passed away. Both of these men were respected firemen and community servants whom Dad and I knew. Farmington also lost Dr. Paul Brinkman who had conducted physical exams for the Farmington Fire Department and Dr. Phillip Chase who had been a family practitioner for years.

Dad was born on June 5, 1922, the only child of Robert Earl McCleery and Adella Luce McCleery. Rob and Della, as they were known, were 43 and 37 years old respectively when Dad was born in the home where he resided almost all his life with exception of the first five years of our parent's marriage. As an only child Dad's mother and aunts catered to him, and Dad said Grampa worried the he might not amount to anything with all the coddling he received. Grampa need not worry as Dad was cut from the same cloth as his parents. Rob and Della were prominent members of Farmington Grange #12. Rob served as a Director of the Farmington Mutual Fire Insurance Company for many years beginning in 1931 where Dad would also serve as Director from 1983 until his death. Rob belonged to the Odd Fellows and Knights of Pythias, was a local representative to the State Department of Agriculture on issues of concern to local farmers, and a member of other farming organizations. Della was a member of several civic organizations and very involved in the Red School House Extension

R.E. McCleery - R.L. McCleery & Adella Luce McCleery

Robert Earl McCleery, Robert Luce McCleery, Adella Luce McCleery
Dad about 1926, age 4

THE LEWISTON DAILY SUN, THURSDAY MORNING, SEPTEMBER 18, 1941

RANDOM SHOTS AT FRANKLIN COUNTY FAIR

Baby Beef entries left to right: Arthur Keyes, Harry Ellsworth, Enoch Bridges, Walter Mosher, Alice Drown, Earl Ellsworth, Eugene Mosher, Stanley Ellsworth, & Robert McCleery with his prize winning steer.

Earl Ellsworth - 4th from right & Robert McCleery - far right.

Courtesy Lewiston Daily Sun, September 18, 1941
L-R: Arthur Keyes, Harry Ellsworth, Enoch Bridges, Walter Mosher, Alice Drown, Earl Ellsworth, Eugene Mosher, Stanley Ellsworth, Robert McCleery, age 19, with prize winning steer

and Red School House Sewing Circle. Grampa was a hard-working, kind, good with numbers, prim and proper man who was devoted to his family and took responsibility for his two unmarried sisters who moved in with our grandparents when they were no longer able to live alone. Grammie was as sharp as a tack and a hard-working, jovial, fun-loving person who loved reading to her grandchildren, sneaking them candy, cooking delicious food and tending her beautiful flower gardens. She was a people person who had a quick wit and great sense of humor.

As a child Dad attended the little Red School House just down the road where his father and aunts had gone to school. Dad liked to play ball, read, draw, and play with the barnyard kittens. He helped his parents on the farm with the Jersey and Guernsey cattle and Oxford Down sheep, and he was a member of 4-H who enjoyed showing his cattle and baby beef at the fair. Dad would say, however, that as an only child living in a houseful of adults it was sometimes lonely not having anyone to play with the nearest friend a mile or so away.

Dad loved history, and he would read anything he could get his hands on about the Civil War. Two of Dad's great uncles fought on opposing sides during the Civil War. Dad's great uncle, Gustavus Augustus Stanley, fought in the Civil War as Captain of the 28th Maine Infantry and 2nd Maine Cavalry. Gustavus was born in Farmington, an older brother to Dad's grandmother, Susan Stanley McCleery. He graduated from Bowdoin College in 1857; studied law in Illinois prior to the Civil War; after the war became a circuit court judge in Tallahassee and Pensacola Florida where he died. John Thomas Stanley, the older brother of Gustavus and Susan, graduated from Bowdoin in 1849; found his way to Texas where he taught at a women's college in Chappell Hill prior to the war; supported the secession of southern states and joined the Confederacy; after the war signed an Amnesty Oath and practiced law in Texas until his death. Gustavus Stanley was given an engraved dress sword by his company when the war ended which eventually passed down to Dad through the estate of his grandmother. His name along with other graduates and students who fought in the War to Maintain the Union is inscribed on a plaque in the lobby of the Maine State Music Theatre at Bowdoin College.

Dad was called "Mac" in high school and starred on the Farmington High School football and baseball teams where he was the center and catcher respectively. Dad even tried his hand at hockey. Dad met Mom who was born on January 1, 1925 when she came to Farmington to attend high school at fourteen. Her father, Albion K. P. Edwards, died when she was ten and Mom and her mother, Sadie Smith Edwards, left the farm on North New Portland Hill so she could attend school in Farmington. Dad and Mom were quite a high school couple according to Janet Mills whose mother, Kay Mills, was one of their teachers and said they would hold hands in her class which she thought was cute.

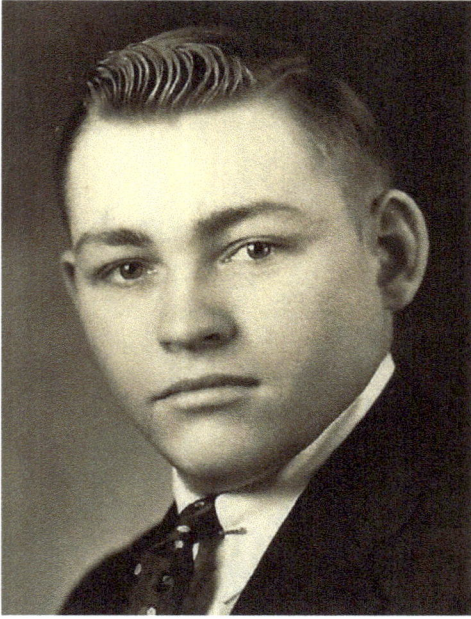

Robert Luce McCleery 1941

Edith Lucille Edwards 1942

Married June 5, 1943
Photos by Luce's Studio

Mom played field hockey, was a cheerleader, and graduated in the top ten of her class. One of Dad's cousins, Arthur Kimball, who was quite a bit younger asked Dad one day while visiting which girl in a group was Dad's girlfriend and according to Arthur, Dad's response was "the pretty one, of course." Dad graduated from high school in 1941 and Mom in 1942. They got married on Dad's birthday, June 5, 1943, and were married 59 years and had six children.

Upon graduation from high school Dad and Grampa ran the farm operation together as partners. It was the R. E. McCleery and R. L. McCleery Farm with at one time 400 acres and 50 plus head of dairy cows. I think it was always expected of Dad that he would take over the farm. Dad certainly had other interests, but his duty to his parents took precedence. Prior to Dad and Mom's marriage Grampa's health was not great, he had suffered with asthma and heart problems for years. Grammie would eventually die from metastatic breast cancer. Five years after Dad and Mom were married Grampa's health had deteriorated to the point where Dad and Mom had to move onto the farm with our grandparents to take care of them and take over the management of the farming operation. Dad got rid of the sheep he didn't especially care for and modernized the farming operation with a tractor and other equipment versus relying on a team of horses to get the job done. In the 1960's Dad and Mom converted the dairy farm into the poultry business until the industry in Maine collapsed in the early 1980's.

Dad's closest childhood friend, Red School House classmate and 4-H buddy, Earl E. Ellsworth, went off to fight in World War II. Earl Ellsworth became a 2nd Lieutenant and served in the U S Army Air Force 33rd Bomber Squadron, 22nd Bomber Group. His B-24J plane went down over the Samar Islands in the Philippines on January 23, 1945, and the plane was not recovered until 1950. This was a difficult loss for Dad, Earl's family, and the community. Dad wanted to enlist, but his parents' health issues demanded otherwise. I believe the conflicted feelings Dad had about this decision shaped the rest of his life.

Dad tried to join the fire service in the 1940's, but he lived outside the Village Corporation limits and therefore could not become a member of the fire department. That did not stop him. Put a challenge in front of him and watch Dad find a way to meet and exceed that challenge. Dad was determined to serve so he found out that if he showed up to assist with a fire call that he would be allowed to work the fire. Dad bought his own gear, would turn out for fire calls, and even beat other firemen to the truck or to the fire scene. It has been told that Dad not only passed a police car to the scene of a fire, but also a fire truck or two. Just as soon as the Village Corporation boundaries were expanded to allow Dad to officially join the department, he did. The rest is history.

Dad and Mom were active in the grange: Dad since 1936 and Mom since 1939. They both served as master of the grange and every other possible position. Dad would perform in one act plays and skits at Grange meetings with Merritt Averill, Orlando Small and our adopted uncle, Don Brown, and as an occasion would have it they would not hesitate to dress up as women with lipstick, scarfs around their heads, and dresses with properly placed accoutrements to the roar of the crowd. Dad was a ham at heart. He loved a good time, a good joke, and enjoyed the camaraderie of his friends and family.

For a social guy like Dad work on the farm could be a solitary business so Dad joined fraternal and other organization available to him: The Masons, The Odd Fellows, The Shriners, The Franklin County Agricultural Society, The Farm Bureau, and Farmington Historical Society. Mom followed suit with her service to the Henderson Memorial Baptist Church, 4-H clubs, Farmington Area Alumni Association and Recognition Committee, Farmington Historical Society, West Farmington Literary Club, Red School House Extension Club and Sewing Circle, and she was the office manager for the Franklin County Soil and Water Conservation District for 21 years to boot.

When asked once about their involvement in nearly every committee, organization or church group imaginable, Mom's response was: "It's not any big deal. We've just tried to do our part." "We always just attempted to make a place as good, or better than when we started. You're not much of a success if you don't." Dad joked and said: "We've been involved in nearly everything that doesn't pay anything."

Dad and Mom lived a purpose filled life. They were loved and respected by many, but none more than their six children, their children's spouses and thirteen grandchildren, whose lives, activities and interests they actively engaged in and supported with every fabric of their being. Dad and Mom were fun loving, and encouraged laughter, togetherness, and family closeness. They set expectations, expected rules to be followed and promises to be kept, and they didn't put up with any foolishness. Dad and Mom were accepting of change and new ways of doing things with few exceptions. Dad's philosophy on life was: "Enjoy life and take it as it comes. There are times when you can change things, and others when you just have to roll with the punches."

Dad's funeral was attended by fire companies from Franklin, Androscoggin, Somerset and Oxford counties, as well fire personnel from as far away as Cape Elizabeth and Freeport, and officials from the State Fire Marshal's Office, forest rangers, ambulance companies, Grangers, Masons, Shriners, Odd Fellows, fair goers, fellow farmers, and friends. Over 300 people came to pay their respects and even more lined the two plus mile funeral processional route; some did not know Dad personally other than what they had just read about in the news coverage of his passing, but they came to pay their respects to what they believed to be a good man who dedicated himself to the service of others, his town, and his state. It was an amazing tribute, and one that

Dad would have been embarrassed by, yet humbled, and deeply moved. As Chaplain Stanley Wheeler said at Dad's funeral: "Bob was highly regarded by those he led and his peers." "Kindness and patience were hallmarks of Bob's personality." "McCleery was a leader among men."

Rest Easy, Dad! You gave it all.

You raise me up, so I can stand on mountains

You raise me up, to walk on stormy seas

I am strong, when I am on your shoulders

You raise me up, to more than I can be

Your Loving Daughter,

~ Ruth

The song *You Raise Me Up* was written by Secret Garden's Rolf Lovland and Brendan Graham and performed in the UK in 2002 and made popular by Josh Groban in 2003; it's an Irish-Norwegian song of which part of the melody resembles the Irish ballad *Londonderry Air*. In 1787, the McCleery family emigrated from Londonderry, Ireland to the District of Maine, part of Massachusetts.

Mom & Dad 10/14/1998

Clyde Ross, Mom, Chris, Will & Ruth Watson 7/26/2002

TOWN OF FARMINGTON, MAINE

147 LOWER MAIN STREET / FARMINGTON, MAINE 04938

(207) 778-6538
PHILIP K. SCHENCK, JR., TOWN MANAGER

January 31, 1977

Robert McCleery
RFD #2
Farmington, Maine 04938

Dear Bob:

I would like to take this opportunity to congratulate you on your
appointment as Fire Chief effective April 1, 1977. The appointment was
made by the impartial review panel set up for interviews on Wednesday,
January 26 and by the Farmington Fire Department at a special meeting
held Friday, January 28. I would urge you to work closely with Forrest
Allen prior to his retirement to learn the duties of your new position.
You should also set up a time with Forrest to select a procedure for
appointing two assistant chiefs to replace Harrison Bragdon and yourself.

Again, congratulations on your appointment, and please feel free to call
at my office, if I can be of any assistance to you.

Sincerely,

Philip K. Schenck, Jr.
Town Manager

PKS/nt

January 31, 1977
Congratulations Letter from Town Manager Phillip Schenck

Your Community in the News

McCleery Is Selected
Farmington Fire Chief

By BARBARA YEATON

FARMINGTON — Robert L. McCleery, Farmington, has been selected to become Farmington's fire chief to replace retiring Farmington Fire Chief Forest Allen, effective April 1.

Town Manager Philip Schenck Jr. announced this week from the town office that McCleery, 54, had been selected by both an oral review panel and the town manager. The panel was composed of Chief Carl McKinney, Skowhegan, and Francis Roderick, State of Maine. The Farmington Fire Department members also approved the selection of McCleery for this post.

McCleery has been with the fire department since 1960 and has been an assistant chief for the past three years. He is presently serving as vice president of the Franklin County area with the Maine State Federation of Firefighters.

McCleery is also active in other town and community affairs. He is past master of Farmington Grange and Excelsior Pomona and a past deputy of the Maine State Grange as well as being a member of the National Grange. He is affiliated with Franklin Lodge, IOOF, and is chaplain of Maine Lodge, AF and AM, and a member of other Masonic bodies and a member of Kora Temple Shrine. He is past president of the Franklin County Shrine Club. He attends the Henderson Memorial Baptist Church.

McCleery is past president of the Franklin County Agricultural Society and is superintendent of the second division, the livestock division of the society. McCleery is a native of Farmington and was educated in local schools. He and his wife, Edith, make their home at the McCleery Farm on the Wilton-Farmington highway and currently operate a poultry ranch. The couple has six children and several grandchildren.

ROBERT McCLEERY

Article Courtesy Lewiston Daily Sun
By Barbara Yeaton
February 5, 1977

TOWN OFFICER'S RECORD OATH AND CERTIFICATE

Municipality of .... Farmington ....

County of .... Franklin ....

State of Maine .... 19 ....

I, ... Robert L. McCleery .... do swear that I will support the Constitution of the United States, and of this State, so long as I shall continue a citizen thereof. So help me God.

I, ... Robert L. McCleery ...., do swear that I will faithfully discharge all the duties incumbent upon me as .... Fire Chief .... according to the Constitution and laws of the State. So help me God.

Signed ...*Robert L. McCleery*

---o---

STATE OF MAINE

.... Franklin .... ss. .... April 1, 19 .. 77

I, ... Fay B. Adams ...., certify that .... Robert L. McCleery .... personally appeared before me on this day and took the above oath.

Fay B. Adams

Clerk
Justice of the Peace
Notary Public
Moderator in open
town meeting

OFFICERS TO BE SEPARATELY SWORN FOR EACH OFFICE
Where the same person is elected to fill more than one office, of which officer an oath is required, he should be sworn separately for each office. For example, in a case where the same persons hold the offices of selectmen and assessors, each person so elected should be sworn (in separate oath) as an Assessor in addition to being sworn as Selectmen. Unless assessors are sworn as such they cannot make a valid assessment.

THIS FORM REPRINTED FROM MAINE ASSESSORS MANUAL BY SPECIAL PERMISSION OF MAINE MUNICIPAL ASSOCIATION.

April 1, 1977
Oath of Office for FFD Chief

TOWN of FARMINGTON, MAINE

OFFICE OF THE TOWN MANAGER
Alan Gove

147 Lower Main Street
Farmington, Maine 04938
(207) 778-6538

November 14, 1983

Fire Chief Robert McCleery
Routes 2 & 4
Farmington, Maine   04938

Dear Chief:

    Congratulations on being elected president of the Maine Fire Chiefs' Association for the year 1984. Election to this position by your piers is indicative of their respect for your interest, and judgement in association matters.

    Please keep me posted on happenings or programs of interest and best of luck on your year as president.

Respectfully,

Alan Gove
Town Manager

AG:njl

cc - Board of Selectmen

Congratulations Letter from Town Manager Alan Gove on
1984 President of Maine Fire Chief Assn.

Meet the McCleery's

These sort of people make the area great

By Karen Kreworuka

Remember when Rtes. 2 & 4 from the Red Schoolhouse Road to the Back Falls Road was all tree-lined fields and open space?

That's the way it was in 1943 when Edith Edwards married Robert McCleery and moved to the strip of highway that was then "all sand."

In the nearly 50 years since then, the McCleery's have seen tremendous changes. They've raised children, given up farming and kept up a steady, impressive involvement in the community, which is "very important" to them, Edith says.

Robert McCleery, ten-year Fire Chief of the Farmington Fire Department, has held leadership posts in the Farmington Grange, the Shrine, the Masons, the Odd Fellows, the Maine Association of Agricultural Fairs, and the State Fair Association. He is past president of the Franklin County Firemen's Association and of the State Firefighters' Association, where he now serves on the Board of Directors.

Edith is past president of the Grange and the Farmington Alumni Association, of which she is now a trustee, as well as trustee of the Henderson Memorial Baptist Church at Farmington, where she will be a 50-year member in December. She was a deaconess, Sunday School and Vacation Bible School teacher, served on the Pulpit Committee, is a former 4-H leader, a member of the Farmington Historical Society and of the Literary Club.

Both Robert and Edith in 1988 were named

Edith and Robert McCleery - (Photo by Janice Daku)

"Outstanding Alumnus" of Farmington High School.

The list goes "on and on," Edith says. But she stops and gives a tour of the huge living room, which is a gallery of family photographs and award certificates.

The McCleery's six children benefited from farm life, but all chose to pursue higher education. All are college graduates, and two, Ruth and Louise, have Master's Degrees.

The hard work continues for Robert, 68, and Edith, 65, who cut their own hay to sell and grow huge vegetable gardens.

They raise as many as 30 varieties of potatoes each

year for the Farmington Grange exhibit at the Franklin County Fair, as well as all of the grains and grasses for that exhibit.

The McCleery's saw change coming to the area 45 years ago when the section of Rtes. 2 and 4 in front of their house was rebuilt.

Before that, they drove 20 milking cows across that road morning and night.

The McCleery's don't have cows any more.

They don't have chickens, either, though they once raised between 32,000 and 37,000 broilers per year. "Complete and very swift changes" came after World War II, Edith says.

"Everything you could possibly think of" suddenly appeared in a "technological whirlwind." The area began to change from an insular farming community when cars became "so improved" that you could drive 500 miles in them instead of five. You didn't depend on just Farmington any more. You had the world.

And Rtes. 2 and 4, new and improved and no longer a sandy lane, began to fill up with side-by-side businesses and a steady swish of traffic.

But Edith is philosophical.

"You have to move with progress," she says, and adjust to the times, though one doesn't "uphold all that goes."

Though not yet ready to retire, she says she is "cutting back" a bit on her full-time schedule.

And Robert, laconic and smiling, dashes off to the Fire Station.

He's too busy to cut back.

Town of Farmington, Maine

147 Lower Main Street, Farmington, Maine 04938
(207) 778-6538

September 9, 1992

Maine Fire Chiefs' Association
Local Government Center
37 Community Drive
Augusta, Maine 04330

Att: Darrel Fournier, President

Dear Darrel:

It is with deep professional respect that I nominate Chief Robert L. McCleery of the Farmington Fire Department for consideration of the first "Fire Chief of the Year". In my twenty-six years plus of local government experience, I have observed many local Fire Chiefs on a daily basis and none fit the honor better than Chief McCleery.

Enclosed is a document that offers some of the high points of Bob's fire career. Also enclosed is copy from a local newspaper that very accurately informs one of his community involvement beyond the fire service.

Deputy Fire Chief, S. Clyde Ross, provided valuable assistance in compiling the information submitted.

Please do not hesitate to contact me if we can be of further assistance.

Sincerely,

John G. Edgerly,
Town Manager

1992 Chief McCleery Nomination Papers Fire Chief of Year

MFCA Fire Chief of the Year
Nomination Form

I/we wish to nominate __ROBERT L. MC CLEERY__ from the town/city

of _____FARMINGTON_____ for Fire Chief of the Year.

__CHIEF MC CLEERY__ meets all of the criteria for the

award. The attached letter describes his/her qualifications and

accomplishments.

Signed: _John B. Edgerly_
        TOWN MANAGER
Address: 147 MAIN STREET
         FARMINGTON, MAINE   04938

Telephone: (207) 778-6538

(Established 1912)

Fire Service Career

Bob, as a young adult was interested in the fire service, but because be lived outside the Fire District, could not become a member of the local Fire Department. When the rules changed to include all those residing within the Town, Bob became a member on February 2, 1960.

Bob served as Clerk of the Department for a number of years and became Assistant Chief for Finance and Administration in February of 1975.

He was appointed Chief of the Farmington Fire Department in March of 1977 and continues to hold that position to this day.

Bob is active in the Maine Federation of Fire Fighters and was elected Vice President representing the Franklin County in 1976 and 1977. He was elected President of the Franklin County Firemen's Association in 1980. During his tenure in office, the Franklin County Attack School was organized and he helped organize a Mutual Aid Agreement in the County. Also during this period, Bob encouraged and implemented and supported fire training especially in the use of SCBA and interdepartmental activity in Franklin County.

In the late 1980's he was a member of the Franklin County Radio Communications Committee which created county-wide dispatch for all emergency services through the Sheriff's Office. Bob continues to serve as Chairman for a regional communications unit that coordinates purchases and service improvements.

Bob is a long standing member of the Maine Fire Chief's Association and is a Past President, Chairman and member of the Fire Education and Training Committee, and is current Director and member of the monimating committee. He is also a member of the International Fire Chief's Association - New England Division.

Bob initiated, encouraged and implemented a rigorous fire department training and education program which continues to stand out in the fire services. The program attracts active participant fire fighters from nearby municipalities. In addition to a strong compliment of certified Firefighter II's, the local department has six certified HAZMAT TECHS, who are members of a regional response team.

Bob is an advocate of having proper and adequate equipment for fire suppression purposes. As chief, he has spearheaded an aggressive long range plan of equipment and gear purchase and replacement.

Major accomplishments during his tenure include:

A. 1982 Mack 1,000 gpm pumper
B. 1988 E-ONE 110' ladder
C. 1989 Fully-equipped Squad Unit
D. New Cascade system that also serves other municipalities.
E. Conversion to positive pressure SCBA units department wide.
F. All personnel fitted with Nomex turnout gear and boots and helmets.

During his tenure an aggressive Fire Prevention and Education program has been developed and implemented. The program serves local schools, University of Maine @ Farmington, industry and the private sector. Components of the program include "Learn Not To Burn" and Maine Pine Tree Burn Foundation materials.

Although one might assume Bob is very serious about the fire service, he does sometimes mix his sociable side into the fire service side. He hosted the 1985 Maine State Federation of Fire Fighters Annual Convention in Farmington and he will repeat that role in 1994.

Local couple receive Grange Public Service Award

The Franklin

AND FARMINGTON ON

A prominent Farmington couple, Robert L. and Edith E. McCleery were presented a public service award, Thursday evening at a meeting of Excelsior Pomona Grange held with Aurora Grange in Strong.

The surprise presentation was made by State Grange Lecturer Clyde G. Berry.

Both Mr. and Mrs. McCleery have a long grange history as well as being active in numerous other civic, fraternal and church organizations.

Mr. McCleery, Farmington Fire Chief, is a Golden Sheaf member of Farmington Grange which he joined on March 31, 1986. Mrs. McCleery joined the grange March 11, 1939. Both have served as master of the grange. He has served in the capacity for seven years, assisted by Edith as Ceres and she was

Local Couple

(Continued From Page One)

degree work over 30 years. Mrs. McCleery went through the offices in Pomona with her husband and has directed Pomona's Court for the fifth degree for over 30 years.

McCleery has been a Maine Deputy eight years and on the grange installing team for several years.

Mrs. McCleery was a Maine State Junior Deputy for six years and also worked on the installing team.

McCleery is past noble grand of Franklin Lodge, IOOF, a past district officer and assisted with degree work. He is a member of Maine Lodge, A. F. and A. M., and is a former chaplain of the lodge. He is past high priest of Franklin Chapter and past illustrious master of Jephthah Council, past commander of Pilgrim Commandery, KT, also a member of Kora Temple Shrine of Lewiston, and has served as a Potentate's Aid and is currently serving as an ambassador.

McCleery has been a member of the Farmington Fire Department 26 years and fire chief six years. He is past president of Franklin County Firemen's Association and past president of the State Firefighters Association and is currently a state director. He is a member of the State Training Committee under the Maine Fire Chiefs Association. McCleery established the Mutual Aid program in the local area, also the local Fire Attack School.

Deputy Fire Chief S. Clyde Ross describes "Bob" as a "go-getter", with high expectations of his people and accountability of his people in the department."

He is past president of the Franklin

master for two years. Mrs. McCleery has also served as grange lecturer.

Robert has been an officer of the grange 39 years and on the executive committee since 1963. Both Robert and Edith have been master of the degree teams and worked on the Farmington Grange Fair Exhibit for over 40 years. She has been chairman of the exhibit committee for over 35 years.

McCleery is past master of Excelsior Pomona Grange, having been a member 50 years. He has assisted with

(Continued To Page Six)

County Agricultural Society, and is currently a trustee and superintendent of the cattle division. He is past president of the State Fair Association and is currently a state director.

McCleery is director of the Farmington Mutual Fire Insurance Company, member of the Farmington Historical Society and Franklin County Farm Bureau.

McCleery raises grasses and grains for the Farmington Grange Fair Exhibit and tags them. He also raises vegetables and specializes in melons for the exhibit.

Mrs. McCleery is a member and former deaconess of the Henderson Memorial Baptist Church. She has served as chairman of the Board of Trustees, is a former Sunday School teacher and former chairman for the Vacation Bible School. She has served on numerous other church committees and is a member and past resident of the Irene R. Luce Philathea Class.

Mrs. McCleery is former advisor of the Happy Workers 4-H Club, is currently serving as a member of the Board of Trustees of the Farmington Area Alumni Asociation and is a past president. She has been active in the work of the Farmington Historical Society and has served as a trustee. She is a member of the West Farmington Literary Club and former member of the Red Schoolhouse Extension group and the Red Schoolhouse Sewing Circle.

Robert Luce McCleery and Edith Edwards were married on June 5, 1943. They now live on the McCleery Farm

where Robert was born. They formerly raised registered Gurnsey cattle but later converted to the broiler business raising up to 36,000 broilers. McCleery does some lumbering and now pastures cattle, a few horses and ponies and manages the family garden with Edith's assistance.

Mrs. McCleery is office manager for the Franklin Soil and Water Conservation District, a position held for 15 years.

She is a member of the Farmington Recognition Committee.

It was interesting to have it noted that the area where Mt. Blue High School is located was sold by the McCleery's to S.A.D. Nine.

The couple has six children and six grandsons and one granddaughter. Their eldest daughter, Susan, Mrs. Robert Small, teaches school in South Portland. Their eldest son, Alton, lives in Farmington with his wife and family and is a machinest inspector for Mid-State Machine Co. in Winslow. Another daughter, Ruth, owns the Consulting and Placement Business, BRW Associates in Portland, where she and her husband reside. David and his wife live in Woburn, Mass. He works in Waltham in the computer consulting business. A daughter, Louise is attending graduate school at the University of Virginia, studying Health Promotion. The youngest daughter, Jane, graduated from Westbrook Junior College and is presently working for her sister, Ruth, in Portland.

Article Courtesy Franklin Journal, July 25, 1986

Maine Fire Chiefs' Association

Local Government Center
60 Community Drive
Augusta, Maine 04330-9486
1-800-452-8786

September 14, 1998

Chief
Congratulations on your
nomination. I'll be at the MMA
Convention on the 14th so I'll
try my best to be there to see
you recognized as a
nominee!
Pam

Robert McCleery
147 Main St.
Farmington, ME 04938

Dear Chief McCleary:

You have been nominated for the Maine Fire Chiefs Association 1998 Fire Chief of the Year award. The selection has been made and will be announced at the MFCA Annual Meeting at the Augusta Civic Center, Wednesday, October 14, 1998, beginning at 9:30 AM.

I encourage you to attend and be recognized. Please wear your dress uniform. If you have any questions, please call Joan at Maine Municipal Association (1-800-452-8786).

I look forward to seeing you on October 14.

Sincerely,

William Page

William Page, President
Maine Fire Chiefs Association

WP:sj

cc: Town of Farmington - Town Manager

(Established 1912)

1998 Chief McCleery Letter from Maine Fire Chief Assn.
Note from Town Manager Pam Corrigan

Robert L. McCleery 213

1998 Maine Fire Chief of the Year 10/14/1998 Pam Corrigan, Farmington Town Manager, Chief Bob McCleery

Members of Farmington Board of Selectmen & Town Managers with Chief McCleery
L-R: Dennis Pike, Town Manager Pamela Corrigan, Emily Floyd, Chief Robert McCleery, Everett Vining,
Former Town Manager John Edgerly, Charles Murray

ROBERT L. McCLEERY
Maine Fire Chief's Association "Chief of the Year"

Farmington Fire Chief Robert L. McCleery and his wife, Edith
(Photo by Sheila McMillan)

There was excitement in the air on October 14, 1998 as the we arrived at the Maine Municipal Association's Annual Convention. For Town officials the event is an annual pilgrimage, but this year was different. Something sneaky was happening and we were loving every minute of it. Dressed in our best, family, friends and co-workers stood vigil at the doorway of our predetermined hiding place to catch the attention of any of our group who may be temporarily confused about the clandestine plan. Once spotted, we quickly ushered any wayward stragglers into the out-of-the-way meeting room so as not to be seen by Chief McCleery who would soon arrive to attend the Maine Fire Chief's annual meeting. Once secretly gathered, our number totalling about thirty, we awaited to be escorted by MMA officials to the in-progress Chief's meeting where officials were about to surprise the unsuspecting Farmington Fire Chief by his selection as "Fire Chief of the Year". We remained unnoticed by Chief McCleery even as we stood in the back of the room filled with fire fighters from around State. The State officials introduced the eleven candidates and then announced the winner. Either Chief McCleery is an accomplished actor or he was truly astounded by the fact that he had been so honored. After a reading of the nomination followed by brief and humble remarks from the Chief, we all retired to our original meeting place where we enjoyed refreshments, hand shaking and picture taking. We were, and continue to be, supremely proud of our Chief and we were delighted that so many of his family and friends were able to join us for the special event.

So that the citizens of Farmington can appreciate Chief McCleery's many accomplishments, we have included his nomination on the following page.

1999 Town Annual Report

NOMINATION FOR MAINE CHIEF'S ASSOCIATION
1998 CHIEF OF THE YEAR AWARD-ROBERT L. McCLEERY

It is with great pleasure that the Farmington Board of Selectmen, Fire Department, and town staff nominate Chief Robert L. McCleery for the Maine Fire Chiefs Association's Chief of the Year. Chief McCleery has selflessly dedicated his life to his profession, his community, Farmington and the surrounding region. His name is synonymous with leader and public servant.

Robert L. McCleery, born June 5, 1922, was raised and educated in Farmington. Robert and his wife, Edith Edwards McCleery, established their home on a farm in Farmington where they raised six children. During the Chief's 76 years in Farmington, he contributed immeasurably to the community. He began community service at a young age when he joined the Grange over 60 years ago and served that organization by holding numerous official positions. Chief McCleery is a York Rite and Scottish Rite Mason, a Shriner, a member of the Independent Order of Odd Fellows (IOOF), past president and remains a very active member of the Maine Fair Association, past president of the Franklin County Agricultural Society, and is the current chairman of the Society's Board of Trustees. In his early years, Robert had longed to be a fireman but lived just beyond the boundary limits of the Farmington Village Corporation. The membership restriction did not stop him from joining firefighters at the scene to offer his support. In 1960, when residency requirements changed, Robert became a rookie member of Farmington Fire Department #1. He served as a regular member and as secretary prior to becoming Assistant Chief of Administration and Finance in 1975. In April 1977, he was elected as Fire Chief and has continuously held that title to the present day.

During Chief McCleery's 22 years of leadership as chief, many innovations and changes have occurred within the Farmington Fire Department and throughout the State as a result of his dedication to safety and efficiency. The department developed increased "live" fire training, use of SCBA, and was one of the first departments to acquire a Cascade refueling air system. Throughout the years, he consistently insisted on a well trained force and encouraged firefighters to attend the Fire Fighter One and Two Academies. In the early 1980's, Farmington's force had seven State Certified Candidates for instructing in Firefighter One. Chief McCleery supported and brought to Farmington the "Learn Not To Burn" program widely used in the elementary school system and he encouraged his officers to participate in educational programs for civic and fraternal organizations. He was instrumental in obtaining extrication equipment through a Department of Transportation Grant in 1984 and subsequently advocated for and obtained the Hurst "Jaws of Life"in 1997.

Under Chief McCleery's guidance, a regional search and rescue unit was established three years ago to offer Cold Water Rescue, Swift Water Rescue, Ice Rescue, High and Low Angle Rescue, Orienteering Rescue, Snowmobile rescue and numerous other critical skills. He has also provided leadership by supporting the creation of a Hazardous Materials Team partnership between Farmington, International Paper, and mutual aid fire departments. Chief was instrumental in upgrading the apparatus of Farmington Department #1 which now includes three pumpers, an aerial, tanker and squad/rescue vehicles and the most recent purchase, E-ONE- 1995 pumper. During Chief McCleery's term as Franklin County Vice President of the Maine Federation of Firefighters, the Franklin County attack school was created. This training lasted 5 years before local departments began to assume greater responsibility for firefighter training and the academy was discontinued.

Chief McCleery is past president of the Maine Fire Chief's Association, past president of the Franklin County Firemen's Association, former director of the Maine Fire Chief's Association and is currently a member of the International Fire Chief's Association - New England Division, a member of the New England Association of Fire Chiefs and a member of the Maine Federation of Firefighters.

Chief McCleery's extraordinary leadership has brought credit to his department, the region and his profession. His willingness to lead by example and to dedicate endless hours to the enhancement and enrichment of the careers of firefighters throughout the state has been remarkable. Chief McCleery is revered by his department and recognized as a leader by his peers. The quality of Chief McCleery's 38 years of service to the firefighting profession is truly deserving of this year's selection of Fire Chief of the Year.

Farmington Fire Fighters Battle the Blaze at the Fairbanks School
(Photo by Sheila McMillan)

Previous Pages Courtesy Town of Farmington 1999 Annual Report

FRANKLIN COUNTY

Peers honor Farmington's McCleery

FARMINGTON — Robert McCleery, in a surprise ceremony Wednesday, was named 1998 Fire Chief of the Year by the Maine Fire Chiefs Association.

The award was given at the chiefs' annual meeting during the Maine Municipal Association Convention at the Augusta Civic Center.

McCleery has been a firefighter for 38 years and has been Farmington's chief since 1977.

While McCleery knew he had been nominated, the award was secret until it was revealed before a room full of his fellow chiefs and town officials.

"McCleery's extraordinary leadership has brought credit to his department, the region and his profession," Town Manager Pam Corrigan said.

"Chief McCleery is revered by his department and recognized as a leader by his peers," she added.

McCleery, 76, was raised and educated in Farmington. He and his wife, Edith Edwards McCleery, established a farm where they raised six children.

McCleery began community service when he joined the Grange more than 60 years ago and served the organization in official positions. He also is a York Rite and Scottish Rite Mason, a Shriner, a member of the Independent Order of Odd Fellows, past president and active member of the Maine Fair Association, and past president and current chairman of the Franklin County Agricultural Society.

In his early years, McCleery had longed to be a fireman but lived just beyond the boundary limits of the Farmington Village Corp. The membership restriction did not stop him from joining firefighters at the scene to offer his support.

In 1960, when residency requirements changed, he became a rookie member of Farmington Fire Department #1. He served as a regular member and as secretary prior to becoming assistant chief of administration and finance in 1975. In April 1977, he was elected chief and has continued to the title to the present.

During McCleery's 22 years of leadership, many innovations and changes have occurred within the department and throughout the state as a result of his dedication to safety and efficiency.

The department developed increased "live" fire training, and was one of the first departments to acquire a Cascade refueling air system. Throughout the years, he insisted on a well-trained force and encouraged firefighters to attend the Fire Fighter One and Two academies.

McCleery supported and brought to Farmington the "Learn Not To Burn" program widely used in the elementary school system and encouraged his officers to participate in educational programs for the public.

He was instrumental in obtaining extrication equipment through a Department of Transportation grant in 1984, advocating for then obtaining the Hurst "Jaws of Life" hydraulic rescue tool in 1997.

Under McCleery's guidance, a regional search and rescue unit was established. He has also supported creation of a Hazardous Materials Team partnership between Farmington, International Paper, and mutual aid fire departments.

Article Courtesy Lewiston Sun Journal October 15, 1998

Asst. Chief Tim Hardy, Chief Bob McCleery, Deputy Chief Clyde Ross
October 14, 1998

Asst. Chief Tim Hardy with Chief McCleery

Deputy Chief S. Clyde Ross with Chief McCleery

Farmington Fire Dept.

Offices At:
147 Lower Main Street,
Farmington, Maine 04938
207-778-3235

Chief Robert L. McCleery
Deputy Chief Clyde Ross
Deputy Chief Terry Bell
Asst. Chief Tim Hardy

January 18, 2000
Farmington, Me. 04938

To the Town Manager and Selectmen of the
Town of Farmington:

I hereby submit my resignation as Fire Chief and a member of the Farmington Fire Department #1 effective July 1, 2000. After 40 years of service, 23 of them as Fire Chief, it is time for a younger, more energetic and innovative person to lead the Fire Service into the next century. I feel that the Farmington Fire Department is in excellent condition and in good leadership hands at the present time.

Sincerely yours,

Robert L. McCleery

Smoke Detectors Save Lives

January 18, 2000 Chief McCleery Resignation Letter

Fire chief puts out career

After 40 years of chasing fires Farmington's Chief Robert McCleery said he plans to retire.

By SCOTT THISTLE
Franklin County Bureau Chief

FARMINGTON — In his 40 years Chief Robert "Bob" McCleery has seen more fires than he cares to remember.

He's also seen a two-truck department with a handful of volunteers grow into a well-equipped, well-trained and respected fire department.

"It's done a complete about face in that amount of time," McCleery said Friday, after announcing his retirement to the town's board of

The fire department budget for 2000-2001 will likely include a full-time chief's position for town meeting consideration but that may evolve to a public safety manager that would include police, fire and rescue.

selectmen Thursday night. At 77, McCleery said he started thinking about retiring last year, but finally made up his mind and decided 2000 was the year.

"I've thought about this for some time," he said. "I don't know why it should be such a surprise. ... I've had my day. It's time for a younger person and all that stuff to take over." He joined the department in February of 1960 as a volunteer fireman and became its chief in April of 1977, serving in that capacity for 23 years.

His retirement also signals a change in how the department will be managed as town officials contemplate going from a part-time volunteer fire chief to a full-time public safety manager. Town Man-

ager Pamela Corrigan said Friday.

McCleery will stay in his position for six more months to get the town through its budgetary process, she said.

The job of chief has become too involved for one person to manage part-time, Corrigan said. The fire department budget for 2000-2001 will likely include a full-time chief's position for town meeting consideration but that may evolve to a public safety manager that would include police, fire and res-cue. "It's still in a preliminary stage and I think the board wants to take a good look at all the possi-

bilities," Corrigan said.

McCleery joined under Chief Harold Hemingway just after Farmington accepted fire coverage for West Farmington village.

"He knew I was interested and asked me if I wanted to join, I was young and foolish and said 'yes,'" McCleery said.

In 1996 the Maine Fire Chiefs Association voted him fire chief of the year.

"It was an honor I never expect-ed," he said. "I was very humbled by it. It was an honor in itself to be nominated, to say nothing about be chosen."

Over his career he has seen a number of fatal fires and but has al-so seen fire-fighting science and equipment evolve and improve. Some fires where lives and build-ings were lost may have been stopped or even prevented, with to-day's methods and equipment.

"I can think of several fires where, if we had the equipment we have today, when we left, there wouldn't be just a cellar hole," he said.

He has seen the price of fire trucks go from about $8,000 to $240,000 and the pay for firefighters go from $1.50 an hour for recruits

to $7. "Which is still not enough," he said.

In the last 10 years he's seen training requirements intensify and records-keeping become more thorough and complete. "We keep records of everything — everything," he said. "Fire calls, injuries, train-ing — everything."

He said equipment and tech-niques continue to improve and evolve. "I tell these guys today they won't even recognize the gear that they will be using 10 years from now," he said. "That's how far it has come and where it's go-ing."

FIREFIGHTER: Farmington Fire Chief Robert "Bob" McCleery said he plans to retire from the department, he's served for 40 years. "It's time for a younger person and all that stuff to take over." He will stay on for the next six months. His departure may signal a change in the way the town manages its public safety departments, officials have said.

Farmington Fire Dept.

Offices At:
147 Lower Main Street,
Farmington, Maine 04938
207-778-3235

Chief Robert L. McCleery
Deputy Chief Clyde Ross
Deputy Chief Terry Bell
Asst. Chief Tim Hardy

April 25, 2000

Town Manager Corrigan:
    Re: Terry S. Bell, Sr.

      I have known Terry S. Bell, Sr. for 24years and have found him to be a very dedicated and conscientious fireman. Terry Bell was the first person appointed a fireman after I became Fire Chief on April 1, 1977. Terry was also the first member of the Farmington Fire Dept. to graduate from the Fire Fighter I Academy. He later became a FF II.

      Terry was appointed Lieutenant in 1983 and moved to Assistant Chief in charge of pay roll and man-hours time keeper. In these positions he demonstrated his ability to get along with the men and others that dealt with department matters. In August of 1991 Terry was promoted to Deputy Chief, a position he now holds. In this position he has been responsible for overseeing the purchase of radios, pagers, turn out gear for the members and breathing apparatus (SCBA). He has kept records on all the items purchased, sellers and prices of materials obtained. This has made our record keeping more thorough and it has help in meeting current standards required by State regulations.

      Terry served as secretary for the Benevolent Association for several years, a task that required good records and adequate correspondences. He continues to be active in the projects undertaken by the association.

      Terry has served Farmington as the local EMA Director, a job that required much dedication and the ability to work efficiently with other departments in times of crisis. He is involved with the Hazardous Materials (Haz Mat) as a qualified Technician. This is a position that involves a variety of tasks at an emergency scene.

      I believe Terry Bell, Sr. is well qualified to be Chief of the Farmington Fire Department.

Sincerely yours,

Robert L. McCleery, Fire Chief

Smoke Detectors Save Lives

April 25, 2000
Chief McCleery Recommendation Letter for Terry S. Bell to become Chief

he Franklin Journal

AND FARMINGTON CHRONICLE

Serving you since 1840

Vol. 160 No. 53

50¢

TWICE A WEEK
TUESDAY & FRIDAY

Firefighters honor their retiring chief

By Greg Davis

FARMINGTON - Some 200 friends, family and fellow firefighters attended Thursday's surprise reunion party for Chief Robert L. McCleery.

They were celebrating his more than 40 years with the Farmington Fire Department, the last 23 of which he served as its chief.

The celebration was one part a presentation of honors, and the other part a roast.

At the end of the ceremonies, Chief McCleery quipped, "The good stuff I did with the help of all of the department. The other stuff I did with the help of my good buddies."

McCleery said that without his wife Edith's "help and backing, I wouldn't have gotten very far in anything." With new Chief Terry Bell and his deputies in place, he said he is "leaving the department in excellent hands. We have an excellent chief with good deputies behind him, and the fire department has a sprinkling of young and old."

Bell handed McCleery a presentation white leather chief's helmet, and he received two chief's horns, a gift certificate for a stay at a bed and breakfast in Camden from selectmen, and numerous plaques, certificates of appreciation and gifts from firefighters who came to

see Chief page 10

Robert and Edith McCleery (Photo by Greg Davis)

Chief CONTINUED FROM PAGE 1

celebrate his retirement from across the state.

In the "roast" portion of the festivities, it was mentioned how call men once backed a fire truck into McCleery's truck, and how, at conventions and other outings statewide he seemed to have "frequently celebrated birthdays - some 148 to 167 times" in restaurants.

The chief admitted, "I was a little apprehensive as to what I might have heard tonight. I appreciate the turnout. I couldn't believe the number of people and the distance some of you have traveled," he told the audience.

At the end of his response, he got a standing round of applause and a spontaneous singing of "Happy Birthday!"

Following a buffet dinner at South Dining Hall at the Student Center, University of Maine at Farmington, an EMS Color Guard had started the festivities.

In the audience were McCleery's six adult children and numerous grandchildren and cousins.

Emcee Stephan Bunker kept things moving, frequently in a humorous vein. Deputy Chief Clyde Ross read letters of congratulations from U.S. Congressman John Baldacci and Senators Olympia Snowe and Susan Collins.

Ross, his cousin, stressed that with more than 23 years as chief and over 40 years of dedicated service, McCleery has made "a lasting impression" on the fire department and town and instituted specialized training that is the standard of today.

Rep. Walter Gooley presented a Certificate of Legislative Senti-

Receives Chief's Helmet - Replacing the leather one he had many years ago, Robert McCleery models the presentation helmet he was handed by Chief Terry Bell. (Photo by Greg Davis)

ment from the 119th Legislature, which noted the many years that McCleery was "called out at all hours of the night for a tough job, dangerous work, with a lot of stress involved, including a lot of stress for the spouses of the firefighters."

McCleery was the Maine Fire Chiefs' association's 1998 Fire Chief of the Year, brought the "Learn Not to Burn" program to town; brought extrication equipment and the Hurst "Jaws of Life"

equipment to the department; instituted Firefighter I and II training, cold water, swift, high and low angle rescue training, orienteering and snowmobile rescue training; and hazardous materials training with International Paper Co. and mutual aid agreements with surrounding fire departments.

Others making presentations, both serious and humorous, included:

• Jeff Brackett, New Sharon Fire Department and Maine Federation of Firefighters.

• Lewis Prescott, Franklin County Firefighters Association.

• Phil Nason, Maine Federation of Firefighters and Maine Burn Foundation.

• John Dean, Maine Fire Marshall's Office.

• Rena Leibowitz, Maine Emergency Management Association and the Maine Association of County Fairs (for which McCleery

has long served as state president and a board member).

• Dick Hall, former East Dixfield Fire Chief.

• Brad Crafts, Franklin County Firefighters Association.

• Mac Burdin, Strong Fire Chief.

• Phil McGouldrick, South Portland Fire Chief.

• Sherm Leahy - Maine Fire Training and Education.

• Charles DeGrandpree, Maine Emergency Vehicles.

• Darryl Fournier, Freeport Fire Chief.

• Farmington Fire Department Auxiliary President Sharon Barker and other members.

• Richard Knight, Farmington Firefighters Benevolent Association.

• Farmington Selectmen Beverly Rogers, Charles Murray, Dennis Pike, Stephan Bunker and Town Manager Pamela Corrigan.

Retired Farmington fire chief dies

By BETTY JESPERSEN
Staff Writer

FARMINGTON — Retired Farmington Fire Chief Robert McCleery was an inspiration for his department, a leader in the state fire service and in agriculture, and a devoted grandfather who was happiest surrounded by his large family.

McCleery, a firefighter for 40 years, died Monday morning from a fast-moving and rare form of cancer at Franklin Memorial Hospital. He was 80.

More than 500 people are expected to attend his funeral to be held at 11 a.m. Friday at the Farmington Fairgrounds in honor of his longtime involvement with the Franklin County Agricultural Association. Following the service in the Starbird Building, the casket will be loaded onto Engine 2 — the last truck the town purchased under McCleery's 23 year-tenure as chief — and a procession of fire trucks will pass through town to Riverside Cemetery.

Farmington Fire Chief Terry Bell who joined McCleery's department when he was 19, said his continuing legacy is the value he placed on training.

"Bob definitely made our department the way it is today," Bell said. "He was an easy person to work with and didn't mind sharing what he knew. And he was always open to new ideas," he said.

Deputy Fire Chief Tim Hardy said it was McCleery's fairness and his encouragement to learn new skills that stand out most in his mind.

"He always wanted us to be the best we could be. He was very, very proud of all of us," he said. "If it wasn't for Bob, I wouldn't be where I am today in the fire service. He was a great inspiration."

Freeport Fire Chief Darryl Fournier was a friend and colleague of McCleery for over 20 years and was at FMH on Sunday to say goodbye to his friend. Both men had been presidents of the Maine Fire Chiefs Association and were longtime members of the board of directors.

"I remember Bob for his honesty, his great sense of humor, and the fact he was never afraid to tackle tough issues on the board," Fournier said.

Please see **McCleery,** *B4*

*** Central Maine Newspapers

• McCleery

Continued from B1

Back in the early 1980s, McCleery served on the state fire training education committee that was responsible for setting the direction for firefighter training in Maine.

He was instrumental in bringing organized training to area firefighters and started the Franklin County Fire Attack School at the fairgrounds. And Farmington was the first town to offer a Firefighter I and II state fire academy for adults on weekends and evenings rather than two weeks in Presque Isle in August.

Richard Hall, a Wilton dairy farmer, was a lifelong friend and the two men were livestock superintendents at the Farmington Fair for decades.

"Bob was always on top of things, and he did it in a quiet, orderly manner. But when he was in a group of people, he was often the ringleader in conversation," Hall recalled fondly. "I just saw him three weeks ago at his 80th birthday party and he was in rare form."

Son-in-law Robert Luce of Carrabassett Valley recalled McCleery's contentment, surrounded by his wife, Edith, his children, and 13 grandchildren and four great-grandchildren. "The more people around him, the happier he was. He loved sitting there, in the middle of all this noise and commotion, with a grandchild on his lap."

Article Courtesy Waterville Morning Sentinel July 23, 2002

Sun Journal

FIREMAN: Farmington mourns the loss of McCleery

PROCESSION: A line of dozens of firefighters salute as the casket of retired Farmington Fire Chief Robert L. McCleery goes to its final resting place Friday afternoon in Farmington.

GRIDIRON BRIDGE/SUN JOURNAL

Article Courtesy Lewiston Sun Journal July 27, 2002

Robert L. McCleery 227

Edith McCleery, above, clutches the U.S. flag from atop the casket of her late husband, former Farmington Fire Chief Robert McCleery, on Friday. A parade through downtown honored the 40-year firefighting veteran, who died Monday of cancer. At right, Farmington Engine 2 bearing McCleery's casket passes by his turnout gear.

Staff photos by Fred J. Field

Article Courtesy Portland Press Herald
7/27/2002

Fire chief put to rest

McCleery a 'leader among men'

BY DONNA M. PERRY
Staff Writer

FARMINGTON—An honor guard of firefighters in dress uniform with black tape across their badges saluted retired Farmington Fire Chief Bob McCleery's flag-draped casket Friday as it was wheeled through Starling Hall before being lifted onto the back of Engine 2 for the last ride through town.

No. 2 was the last fire truck the town bought under McCleery's command before he retired in June 2000 after serving 40 years on the department with 23 of those as chief. McCleery, 80, died suddenly Monday from a fast-moving, rare form of cancer.

The soft ringing of the fire bell echoed through the hall on the fairgrounds accompanied by another firefighter salute, before Farmington Fire-Rescue Lt. T.D. Hardy, at the wheel of Engine 2 with Capt. Mike Bell beside him, flipped on the truck's red and white flashing lights. Two firefighters guarded the casket from the front, and two others from the back, as the truck began to roll. Farmington Police Chief Richard Caton III led the funeral procession with current Fire Chief Terry Bell and other officers in front of Engine 2.

Behind the family's vehicles, there was a long line of firetrucks, with lights flashing, from Androscoggin, Franklin, Oxford and Somerset counties and others from Cape Elizabeth and Freeport, as well as officials from the State Fire Marshal's Office, forest rangers, ambulances, Grangers, Masons, Shriners, fair-goers, fellow farmers and friends. People watched as the cortege moved from the fairgrounds to the former fire station on High Street and through

TRUCKS: Farmington Engine 2 carries the casket of McCleery under an arch of ladder trucks as it makes its way to Riverside Cemetary.

GREGORY RICE/SUN JOURNAL

the downtown area to the current fire station on Farmington Falls Road.

Engine 2 stopped in front of the station where McCleery's reddish office chair sat with his helmet and coat in its seat and the rest of his turnout gear, his pants and his boots, placed on the ground in front ready to be stepped into. Firefighters on the sidelines saluted the casket as Hardy blasted Engine 2's horn three times before the truck moved forward under a black-draped arch formed by Farmington and Jay aerial ladder trucks with a red, white and blue banner on top.

The procession continued on up the road turning into the Riverside Cemetery to the burial site. More than 300 people attended the memorial service at Starling Hall where McCleery had spent time as a member of the Franklin County Agricultural Association.

During the service, McCleery's casket was surrounded by memorabilia, including family photos with a backdrop of the department's 1931 Maxim fire truck. McCleery was remembered as "a leader among men," highly regarded by those he led and peers, said Farmington Fire-Rescue Chaplain Stan Wheeler.

"Kindness and patience were hallmarks of Bob's personality," Wheeler said.

He said McCleery chased firetrucks even before he was allowed to join the department to help in anyway he could. Prior to 1960, only those who lived in the Farmington Village Corp. could be on the fire department.

SEE CHIEF PAGE B4

Chief

CONTINUED FROM B1

"It was often a pride to him that he was often there before the fire trucks," Wheeler said, which brought some laughter from those attending.

Farmington Deputy Fire Chief Clyde Ross, who worked under McCleery and also was his cousin, fondly remembered the time they spent working together in the fields doing corn and hay and chasing the white-tailed deer through the field on Mosher Hill. He also noted that he and others in the family looked up to McCleery and his wife, Edith, of 59 years.

"Today marks the end of an era," Ross said.

When McCleery became chief, Ross said, he was full of ideas and had a vision for his

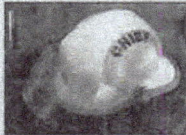

GREGORY RICE/SUN JOURNAL

McCleery's helmet sits on his chair as a tribute to his service

department. But he didn't accomplish it all on his own, he said, "Chief Bob" delegated authority to his officers. Under his watch, fire attack school in Franklin County was started, the Learn Not to Burn Program was introduced to schools and state-recognized firefighter and officer training came to the county among other programs introduced for local fire service.

dperry@sunjournal.com

Article Courtesy Lewiston Sun Journal
July 27, 2002

ROBERT L. McCLEERY
1922–2002

FARMINGTON — Robert L. McCleery, 80, of West Farmington, died early Monday morning July 22, at Franklin Memorial Hospital.

He was born June 5, 1922, in Farmington, the son of Robert E. and Adelia (Luce) McCleery. He graduated in 1937 from the Little Red Schoolhouse and in 1941 from Farmington High School.

On June 5, 1943, he married Edith L. Edwards. Mr. McCleery, a native son of Farmington, will long be remembered for serving his community well.

He was a dedicated member of the Farmington Fire Department for 40 years, serving as Chief for 23 years. He was Vice-President of the Franklin County Fire Association, President of the Maine Fire Chiefs Association and was a member of the Maine and National Fire Chiefs Association. In 1998 he was honored as Maine Fire Chief of the Year. For 33 years, he was a Trustee of the Franklin Country Agricultural Society, serving as Chairman for many years. He was Past-President of the State Fair Association and served on the Legislative Committee of the Maine State Fair Association.

He was a member of the Royal Order of Masons, Maine Lodge 20 A.F. & A.M of Farmington, and served the Jephthat Council for over 25 years, as well as Franklin Chapter and held top office in both the Council and Chapter. He served as Commander of Pilgrim Commandery No. 19. He was a member of Kora Temple Shrine: Past Potentate's Aide under Potentate George Berry, Past-President of the Franklin County Shrine Club and served on the Masonic Building Committee.

He was also a member of the Odd Fellows Society for almost 60 years and served as Past Grand Master and District Warden. For 66 years he was a member of the Farmington Grange 12. He served as Past Master of Farmington Grange, Past Master Excelsior Pomona, past Deputy, Maine State Grange for five years (Excelsior and New Century Pomona). In 1980, he received the Excelsior Pomona Award for Public Service.

He served on the Budget Committee for the Town of Farmington and for three years served on the Farmers Home Assoc. Board. He was a past member of the Farm Bureau; Member of the Farmington Historical Society and Director of the Farmington Mutual Fire Ins. Co. He was honored in 1966 as Outstanding Alumnus of Farmington High School, along with his wife, Edith.

He had a keen interest in history of the Civil War as his great-uncle, G. A. Stanley, served as Captain of the 28th Maine in the war. He supported his family by raising cattle, chickens and other livestock on his farm on the Wilton Road in West Farmington. He farmed actively, even baling hay as late as June 2002. His children followed at his feet as he did the farm chores and field work. He was very patient with them.

He enjoyed life because he lived it well everyday. He lived it with his family, with his friends and co-workers. He lived it in the Grange, Odd Fellows and all the Masonic orders. He was dedicated to the Temple and the Shrine. He was drawn to the Fire Department as though by a strong magnet and he worked his way to the top locally and at the County and State level. He was the first Fire Chief to be elected from a volunteer department.

His dedication to the training and safety of his men was tremendous. He made the department one of the finest in the State for its size through good safe equipment and excellent training. His service to the fire department and to the Agricultural Fair has made him countless friends across this State and many deem it an honor to call him "brother."

Many, many hours have been voluntary and he never regretted his time spent. Other than his various interests in organizations, his hobby was husbandry. His hobby was growing of grains and grasses, potatoes, and numerous varieties of vegetables that were exhibited in the Farmington Grange Fair booth. His trellis of grains and grasses were exquisite. He loved sports and always supported his children's teams as well as others. He was a proud Mainer and a proud American.

He is survived by his loving wife of 59 years, Edith of West Farmington; six children, Susan M. Small and her husband, Robert A. Small of South Portland, Alton E. McCleery and his wife, Cheryl of Farmington, Ruth E. Watson and her husband, William Watson of Alexandria, Va., David R. McCleery and his wife, Susan of Woburn, Mass., Louise A. McCleery of Hampton, N.H., and Jane S. Luce and her husband, Robert Luce of Carrabasset Valley; 13 grandchildren, James McCleery, William McCleery, Tricia McCleery, Robert Small, Steven Small, Christopher Watson, William Watson, Lane Watson, Heath Watson, Erica Luce, Emily Luce, Erin Luce and Elise Luce; and four great-grandchildren, James Alton McCleery, Jacob Andrew McCleery, Malachi James McCleery and Collin Levi McCleery.

Obituary as Published in Lewiston Sun Journal
July 23, 2002

APPENDIX

3 CENTS PER COPY.

Wilton Record.

EXTRA.

VOL. VI.   WILTON, MAINE, OCTOBER 27, 1886.   NO. 274.

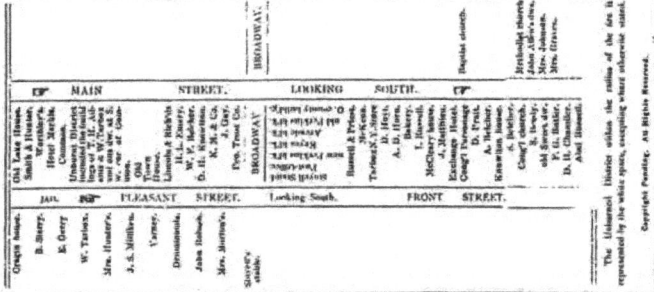

Outline Map, Showing the Location of the Ruins.

$300,000 FIRE AT FARMINGTON!

A Large Portion of our Sister Town Gone.

Including Three Churches, Post-Office, Three Hotels, Forty-Two Business Concerns, and Thirty-Two Dwellings and Stables Connected.

SAD HAVOC ON EVERY HAND.

Ninety-Six Families Homeless. Both Newspaper Establishments in the Same Condition the Record Office was in last February.

Last Friday afternoon at about 3 o'clock, flames were simultaneously discovered by the construction crew of the railroad and by Capt. E. I. Merrill's son Dana, burning from a barn owned by the Stoyell heirs, situated on the west side of Front street, near the narrow-gauge railroad track. An alarm was immediately given, and in a few minutes nearly all the population was at work trying to extinguish the fire. The village has only a small hand fire engine and a hook and ladder company, both of which did good execution, considering the scarcity of water at hand, and at 4 o'clock the fire at the barn seemed to be entirely under control, although a strong guard was placed around it. At this time it was generally thought that the danger was over, but it was not so ordained. The barn contained 45 tons of fine hay, farming tools, etc., and the sparks from the fire were carried in all directions, igniting the shingles of the adjoining wooden buildings, after having been gently fanned by the zephyrs that at times held high carnival. And at about 4.30 o'clock, as Albert Sterry was

standing in his yard he saw a light in E. Gerry's stable, and on inspection the building was all ablaze on the inside. Again the alarm was sounded and caught up by every alarm in town, and in a second all the people were on a "wire edge", and no effort was spared to control the fiery monster; but the efforts were of little effect, for it spread like wild fire in every direction, and all the rest of the night things were sad to behold. Buildings were burning and falling on every hand, and men, women and children were mad with excitement, which is not at all surprising to a party who has ever experienced a burn-out. Furniture and valuable household goods were scattered to the winds, and unnumbered thieves seemed like ravenous wolves seeking to destroy what the fire should leave behind. It is said that thousands of dollars' worth of valuables were stolen. It does seem awful that at such times the land sharks will stoop to such meanness as to pilfer from penniless and homeless people. It may be wrong, but we do not think "lynch law" is too good for the perpetrators of such heinous crimes.

At 9 o'clock Phillips was telephoned for aid, and at 10.30 o'clock Phillips' hand engine and hook and ladder company and 200 citizens from that place were in Farmington, having been drawn the 18 miles in just one hour over the little railroad. We are told that they did considerable execution with their fire department. And no doubt Farmington is very grateful to them.

Lewiston and Portland were telegraphed, for aid, and generously responded by immediately sending fire engines, and it is due to the good execution of them that many more sets of buildings were not burned. Portland and Lewiston will never be forgotten by Farmingtonians. The county jail was consumed, but by remarkable forethought the turnkey, H. Jewel freed the incarcerated prisoners, and thus saved them from a terrible death. The old express horse and a cow that were in E. Gerry's stable were burned. These were the only animals things that got singed that we know of, and it seems truly miraculous that it

is so. It was said that the entire village seemed to be in a rain of fire, as if the fire was coming down from the clouds. Almost every house and store on Main St. were stripped of their contents and great damage was the result, where no fire frequented. Farmington never saw such a season and it certainly does not wish to do so again. Another like conflagration would wipe the village from existence. Sad are the hearts of all her residents now.

By consultation of the map herewith published an accurate idea can be obtained of the location of each dwelling house, store, or building consumed. To one knowing the former appearance of the village, a glance at the map will tell them the story in an instant.

We have been on the ground and taken notes, and, although hurriedly written, we believe that below is as accurate an account as can possibly be given at the present time, of the buildings burned, and in many cases the amount of the losses together with the insurance carried by the owners of the property. It is estimated that the dwelling houses destroyed had an average value of $2,000 each, and the stores $3,000 each. The three churches were valued at about $20,000; insured about one-third. Both of the printing offices bit the dust, and the short town of the county does not possess enough type to print a hand-bill. We truly sympathize with them, and our office and materials are at their service. No one can appreciate a burned up printing office better than we. It cannot be called anything less than discouraging.

The only buildings on the west side of Main St., from Amos Dolbier's on the north, to the late Edwin N. Stevens' house on the south, that now remain standing, are: One dwelling at the south-west corner of the Common, the stores of T. H. Adams, the Tarbox building adjoining, the Exchange hotel stable, and the Free Will Baptist Church, the two latter of which were on the back side the opposite lots and were protected by the ample blowing from them. A pretty narrow escape. Six sets of

buildings on the east side of Main St., including the First Baptist Church and the Methodist Church, the best church edifice in the county, were consumed also.

Beginning at the old Lake House, recently refitted for a dwelling house and occupied by J. M. S. Hunter and James Smith, we proceed south. Before the above parties last about every thing. M. L. Worthley's loss about $1700. Ins. $800. Hotel Marble, furniture $1,000, buildings $5,000. Across Main St. from the store buildings Currier Tarbox, E. O. Greenleaf, L. G. Preston, Mrs. Norcross, Marchant Holley, Ed. Richards, and Cal. Carville's, household goods, were used very roughly and considerably damaged, their houses all having been pretty well cleared. Geo. Cragin's house was blistered by the heat. The trees in front of Hotel Marble on the Common were badly blackened and it is feared that they are mostly dead. Now let's step off to the west to Pleasant St. First, the county jail; west of that, the old Cragin dwelling and buildings connected. Mrs. Cragin's loss is about $1,500. Small insurance. Every house on the west side of Pleasant St. was con-

(Continued on next page).

We shall issue the "outline map" published at the top of this page, showing the location of the ruined buildings, on a neat card, and they can be obtained for the trifle of five cents each. This map is a very accurate one, and will no doubt be very handy as a reference, to jog the memory of parties interested at any future time. For old residents of Farmington who are unable to view the ruins with their own eyes, it will be found an invaluable aid in giving them an accurate idea of what the fire consumed. The cash coming from the sale of the "outline maps"—over and above the actual expense of publishing them—will be given to the poor and needy of Farmington's sufferers. Send your orders to R. A. Merrow, editor Record, or leave them with Fred S. Merrow, Farmington, and they will be faithfully filled.

sumed. Going south E. Gerry's buildings, the Hunter buildings, Varney's blacksmith shop, John Robash's buildings, Mrs. A. M. Mortons buildings. Her loss is $1000. No insurance. On the East side of Pleasant St. stood the old town house which "went up" with the rest. Below Mrs. Mortons plow factory was the stable where the first fire started. Now we have reached Broadway, and proceed East to the square: First was the Stoyell heirs elegant estate. Next came the new "fire proof" block of F. C. Perkins, occupied by Postmaster Merrill, E. G. Blake, jeweler, and Mrs. Wm. Randall, dressmaker. F. C. Perkins loss in this fire is $12,000. No insurance. E. G. Blake's loss, probably, $5,000. Insured $2,250. Next came the *Chronicle* office, which was valued at $10,000. Insurance only $2,-000. Total loss, excepting the books, which were saved by herculean efforts. Over the *Chronicle* was Judge Chandler's law office and Municiple Court room. Arcade block came next. It was owned by Mrs. A. R. Phinney, who loses $8,000 by this fire, and only had $2,700 insurance. This block was occupied on the first floor by A. S. Butterfield, boots and shoes, whose loss is not far from $4,000. Partially insured. Starbird, the photographer, occupied the upper part of this building as a studio. Then came the old Perkins block, in which was located L. A. Smith's Music and Confectionery store. His loss is $3,500. Insured for $1,200 only. Also the store of Ed. Richards, variety and tobacco store, and Lewis Burbank's jewelery store; No insurance. Over these stores, we believe was located E. O. Greenleaf's law office. Mr. Greenleaf's loss is about $2,000. Insurance $200.

Now we have reached the old county building on the corner of Main St. and Broadway, which was occupied by Geo. McL. Presson, jeweler, and Z. H. Greenwood, insurance broker. Mr. Presson's loss is probably $1,000 over and above his insurance. Now we will proceed north. On the corner, west side of Main St., was located the People's Trust Co. and Jabez Gay, boots and shoes, on the first floor of Boardman block; and over them was the law office of Jos. C. Holman, Co. Atty., and Geo. Currier, insurance broker. Over them in the same block the Odd Fellows were comfortably located. The valuables of the Trust Co. were in the fire-proof vault and are supposed to be safe; but their elegant furniture and office fixings were destroyed. Mr. Gay's loss is heavy, and we believe Mr. Holman's is also. Now we reach the Knowlton block, in which were located the printing office materials and fixings of Knowlton, McLeary & Co., the *Journal* Newspaper Co., D. H. Knowlton's news stand and book store, and the Boston Clothing House, W. F. Belcher. Knowlton, McLeary & Co.'s loss is estimated at $15,000 to $20,000; insured for $8,000. Mr. Knowlton and the Newspaper Co. lose heavily; partially insured. W. F. Belcher's loss is probably $7,000; fully insured. Then came H. L. Emery's variety store. Emery's loss is $3,500; insured $1,500. Lincoln & Richards' dry goods store occupied the next store north, the second floor of which they used for a tailoring establishment. Over these stores was the nicely fitted-up home of the Free Masons. It seems that church and society fared alike in this conflagration. No partiality was shown.

Again we will start at the jewelry store of Russell & Priest, the first store south of the old county building on the west side of Main St., and go south. Russell and Priest's loss is $2,000; no insurance. Next south of them was the newly fitted up oyster house of McKeon. Over these stores were one or two tenements, in one of which Mrs. Taylor lived. Her loss is heavy. Next was the New York store of Tarbox Bros. It was a total loss; insurance $500. Next was Daniel Hoyt's dwelling and buildings. His loss was probably about $3,000. Insured $1,200. A. D. Horn's loss was about $1,500. Small insurance. Adjoining Horn's building was the Bakery and Confectionery store. Next south, the residence of Isaac Russell. Then came the residences, dwellings, etc., in the order in which they are named:—The McCleary homestead, occupied by several families, among them Walter Nichols; Jos. Matthews's new set of buildings. His loss was about $3,000; insured $1,000 on furniture. Exchange Hotel, occupied by S. Knowlton, loss on building $2,500. Mr. Knowlton's loss on furniture, etc., was about $1,000. Insured for $800. Congregationalist parsonage, occupied by Elder Pope; the Daniel Pratt dwelling; the A. Belcher dwelling; the Knowlton dwelling; the Sarah Belcher dwelling; the Congregational Church; the S. Beedy dwelling; the "Sweet" dwelling; the F. G. Butler homestead; the D. H. Chandler dwelling, insured $2,000, loss $1,500; Abel Russ[illegible] stable and el., and might as well taken the balance of the buildings as far as value is concerned, insurance, $2,000.

Now we will cross to the East side of Main St. and go North. First was the home of Mrs. Graves, then Mrs. Johnson's, then Campmeeting John Allen's two dwelling houses, one occupied by Mrs. Chas. Norton. Mr. Allen's loss is $1,500. Insured $1,000. Now comes the M. E. Church recently built at an expense of $15,000. Ins. $7,000. And last, was the First Baptist Church, which was off by itself at least twenty rods from the nearest fire. It seems very strange that this edifice should burn. The houses all along the street were somewhat marred, and the home of Dr. J. J. Linscott particularly looked as if it could not have stood the intense heat much longer.

Farmington's conflagration has caused intense excitement throughout Franklin County, and large crowds have been pouring into town from every direction to examine the ruins.

FIRE POINTS.

The post-office is now located at F. A. Davis's furniture rooms on Broadway.

Arthur Keep lost his room furniture. He roomed over the Boston Clothing House.

Dr. H. M. Hard[illegible] lost about $200 worth of goods that was in his room at Hotel Marble.

L. J. Lyons, the Red Store man, reports a loss of [illegible] to $5,000, in goods destroyed and stolen.

Geo. Titcomb's variety store was used roughly, as was also all the others on the east side of Main St.

Post-Master E. I. Merrill saved all the mails, records and other government property of any real value.

A. D. Horn's horses and carriages and equipments were saved and temporarily located in the stalls at Franklin Park.

32 dwelling houses, and 42 business concerns, succumbed to the flames, and 98 families (it is said), are reported as turned out doors.

Zina Greenwood's Ins. Agencies had risks that suffered to the extent of $70,-000; Geo. Currier's Cos., $48,000; D. V. B. Ormsby's Cos., $20,000; and Flo. Voter's Cos., $10,000; giving a total of $148,500, of insurance that we know was held in policies.

A special train from Lewiston, brought three loaded cars of spectators to look over the ruins Saturday forenoon. Trains were also run down the narrow-gauge roads, and were crowded with people.

Frank Gordon hauled the hose-cart of the Portland Fire Department from the depot up the hill on to Main St. with a little horse that weighed only 800 lbs. That was good work for a pony.

Miss Lillian Ramsdell, from Milo, a Normal student, who resided at E. Gerry's, lost everything she had, excepting the clothes she had on. Among her other worldly effects was $35 in money.

It is said that E. Gerry was so badly burned and injured the night of the fire, that his life is yet despaired of. Also that a young man by the name of Hardy has become insane from effects of the fire.

The report about a large fire at Skowhegan, probably came from the following dispatch dated at Skowhegan, Oct. 23: The house and barn owned by Eastman Hathorne in Athens, one mile west of the village, burned Friday night. Loss $1,200; no insurance. The fire is said to have been incendiary.

The report that was rife Saturday about a large fire at Brunswick originated from the following bit of news: Quite an extensive baker's shop on Mill street, Brunswick, owned by the Canadians, was totally destroyed by fire, Friday evening. The hydrants of the Pejepscot Water Co., were used for the first time, and proved very efficient.

NARROW ESCAPE FROM DROWNING.—Last Saturday afternoon, while out duck-shooting on Wilson Lake at about 5.30 o'clock P.M., Prof. I. C. Phillips of the Wilton Academy, and Mr. M. Douglass, a rising young lawyer from Dakota, formerly of Dixfield, were capsized when near the middle of the lake. They saw a number of ducks and both discharged their guns at about the same time. Mr. Phillips' gun, "a little mulish," kicked quite severely, causing him to lurch backwards and jump slightly, which action caused the boat—a small flat bottom skiff and not very seaworthy—to turn over, thus precipitating its occupants into the cold and cruel waves. Having remarkable presence of mind, they managed in some way to get hold of the boat after it had rolled over and around several times, and Mr. Phillips got his arm up over the bow of it, while Mr. Douglass, whose gun was left in the bottom of the boat, seized it and lay it cross ways of the stern end and then grasped the ends that extended out on either side of the boat. They were on opposite sides and opposite ends of the skiff and in that way managed to keep the balance of it and float until assistance came. And in that position, strung up in a very similar way to the pigs in a large abbatoir on gambrel sticks, they hung and took turns hallooing, yelling and calling for assistance for nearly an hour before it was known what the trouble was. Finally their cries were heard, and several men, among whom were Geo. Holbrook, N. W. Sewall, Oris Miller, Chas. Vaughan and Stephen Bonney, took boats and went to the rescue, and by the aid of lanterns and the cries of the drowning men succeeded in finding and rescuing them from a watery grave—none too early, however, for they were both considerably filled with water and completely exhausted and surely could not have hung on to the boat many minutes more. The water was very cold, and it seems almost a miracle that they had strength to hold on as long as they did. As soon as they were fairly into the boats they took them directly to Mr. Holbrook's house, and by the time the exhausted men were got there Drs. Adams and Young, who had been informed of the catastrophe, were on the spot also, and rendered timely assistance. In a few hours, the doctors, assisted by Mrs. Holbrook and the ladies and others present, succeeded in resuscitating their patients so far that they were enabled to have warm dry woolen under flannels put on and were laid side by side in a comfortable bed. Sunday morning they were so far recovered that they were removed to Mr. Phillips' home, where they are fast gaining strength. We pronounce it a miraculous escape from premature death. Mr. Douglass and Mr. Phillips were classmates at Bates College together. Mr. Douglass has lived at Columbus, Dakota, for several years; but is soon to locate at Duluth, Dakota.

MR. SIMON COLLINS KILLED.

Latest News.

Judge D. H. Chandler will probably hold municipal court in the Probate Court Room for the present.

The People's Trust Co. have begun business again, being now located in the new County building. Their vault was being opened to-day.

Knowlton, McLeary & Co., for temporary purposes, have leased rooms with Col. G. H. Wells, at W. Farmington, where they will have one or two presses running in a week or two. Business office for the time being at F. E. McLeary's residence on Main St., Farmington. They show the right metal, and we wish them speedy recovery from their unfortunate condition.

The sheriffs have arrested a chap who gives his name as Gill Fey; but who was released from the State Prison at Thomaston, the 5th inst., and was registered there as James Stevens, and he is now in custody, suspected of being a Farmington incendiary. He is a young man and wears a downcast and melancholly look, apparently not in a fit mood to converse intelligently.

This (Tuesday) morning, at about 9 o'clock, as Simon Collins, a respected citizen of Farmington, stone cutter by trade, was wheeling some lumber on a wheel-barrow in the road opposite the depot on Front St., a runaway horse belonging to D. W. Austin and attached to a carriage came crashing down upon him and he was instantly killed. His daughter, on hearing of the sad accident, fainted, and her life was despaired of.

A citizens meeting was called Sunday, and a committee on "Ways and Means" of 24, with I. Warren Merrill as chairman, was chosen to look after donations and subscriptions, and see to the proper use of everything received; also to have an eye to the necessity of placing proper and respectable looking buildings on the lots made vacant by the fire. The three churches will be rebuilt; and the *Chronicle* will also have a new office and materials very soon.

An associate press dispatch to the *Lewiston Journal* dated at Portland, Oct. 20, says: The case of Geo. W. Goldsmith, a missing boy, on which State Detective True has been at work for some time, is one of the most puzzling he ever had. On the 24th of last July, young Goldsmith, an auburn-haired lad of 18 years, left the home of his father, Joel W. Goldsmith, and all efforts to find him have been unsuccessful. His parents are respectable people and are deeply grieved at his absence. The boy had been at work in the woolen factory at East Wilton, and upon investigation, Detective True found that one of the hands employed in the factory the past summer, was one of the famous "Briggs boys," a notorious trio of burglars, who have plied their trade in nearly every state in New England and done time in nearly all its prisons. One of the brothers is now serving a fourteen years sentence at Charleston. They are crafty, smart, young fellows, who flinch at no kind of deviltry. The one who made a stay at Wilton disappeared at about the same time the Goldsmith boy left, and the detective is satisfied Briggs enticed the lad away for the purpose of initiating him into the secrets and life of the professional crook. But, since his flight, only one clue to the lad's whereabouts has been obtained. That was a letter dated at Boston, received by his father from him, a few weeks ago. The letter was written in an affectionate spirit, but was a very blind composition. The boy intimated that he was in bad company, was sorry for it, but was not able to get out of his predicament, just then. The boy's father and the detective suspect foul play.

Plan Holiday Fete

Delivering the Christmas tree for Farmington Fire Company's annual holiday fete for third graders 6.30 p. m., Wednesday at the Central Fire Station were, left to right, John Bell, chairman; Assistant Chief Maurice Taylor, and George Yeaton. Other committee members are Chief J. Bauer Small, Peter Durrell, Harold Hemingway, Assistant Chief Forrest Allen, Richard Russell, Robert McCleary, George Hobbs and Clifford Neil. (Sentinel Photo by Maguire)

Article Courtesy Waterville Morning Sentinel,
Photo by E. T. Maguire
December 18, 1962
Getting ready for Annual Christmas Party at Station
L-R: John (Jack) Bell, Assistant Chief Maurice Taylor, George Yeaton

Photo by Barbara Yeaton Open House at Station for Fire Prevention Week October 12, 1977
Pictured: Unidentified Woman, Chief Robert McCleery, Howard White

REPORT OF THE FIRE STATION STUDY COMMITTEE

To the Citizens of Farmington:

At the 1973 annual town meeting, the voters of Farmington felt that consideration and attention should be given to the current status and future needs of our fire department. Therefore, they authorized the Board of Selectmen to create a Fire Station Study Committee to determine and make recommendations on the physical needs of the Farmington Fire Department.

The Selectmen appointed the following committee members: Harold Bean, Lewis Fitch, Don Fletcher, Bauer Small, Harvey Smith, Roger Spear, and Robert Verhoeven. Roger Spear was elected chairman and Robert Verhoeven secretary.

During this past year, the Committee was very active, and held a total of ten meetings. One meeting consisted of touring the fire station and municipal office complexes in the Towns of Winslow and Gardiner. Another meeting consisted of spending a full day touring Farmington with a consultant, and former Fire Chief of Auburn, who has over 20 years experience in fire prevention and firefighting.

Based on our consultant's findings and suggestions, our visits to other facilities, our discussions with officials of other towns, our knowledge of the Farmington Fire Department and its present facilities, and additional committee discussion, the Committee is making two recommendations to the voters of Farmington which are in the form of Articles to be voted upon in this year's annual town meeting.

The first recommendation requests that approval be given to the concept of constructing a new municipal building for the fire department, police station, and town officials. With this approval the Town would commit $7,000 to be paid to an architectural firm to develop schematic plans and construction cost estimates. An additional $500. would be appropriated to obtain an option to purchase a building site. No site selection has been made to date. The construction plans and proposed land purchase, once developed, would be presented to the Town at a future town meeting. At that time, the citizens of Farmington would approve or disapprove the building and land acquisition. The Committee is recommending that all municipal services be combined in one building.

The Committee feels that town government and related services can perform most efficiently when such activities are closely coordinated. Coordination can best be accomplished when all activities are housed under one roof. The physical deficiencies of the present town offices were also a major factor for our recommended total municipal services complex.

The present fire station, which is rented from the Farmington Water District, has deficiencies that create current problems and which severely limits future growth. The downtown location presents traffic problems and thus slows response time to fires. Adequate parking is not available for firemen when called to a fire. The present station cannot provide for future growth of the department. An aerial or snorkel truck is a future necessity which could only be housed in a new fire station. Full-time firemen will someday become a reality and space for quarters must be available. Without overcoming these deficiencies and allowing for future growth and improvements, the town's fire insurance rating will probably drop from a "C" rating, to a "D" and possibly to an "E" rating.

Our second recommendation is to appropriate $24,000 from 1973 federal revenue sharing funds for the Fire Station Construction Reserve Fund. At the last annual town·meeting, the citizens of Farmington recognized the potential need for a new fire station by creating a fire station construction reserve fund in the amount of $10,000. Our recommendation would increase the fund to $34,000.

Many factors were considered by the committee during the past year relative to our assigned duties and ultimate recommendations. All recommendations were based upon the Committee's desire to build a fire department which can best serve the needs of the community in its responsibility to protect life and property from fire.

Respectfully submitted,

ROGER G. SPEAR, Chairman
ROBERT VERHOEVEN, Secretary   LEWIS FITCH
HAROLD BEAN   BAUER SMALL
DONALD FLETCHER   HARVEY SMITH

Fireman Harvey Smith & former Chief Bauer Small represent FFD
Courtesy Town Farmington Annual Report

PREPARING FOR DEDICATION — Workmen were busy late Friday afternoon trying to make the grounds in front of the new $600,000 Municipal Building and Fire Station on Lower Main Street ready for a dedication and flag-raising ceremony at noon on Saturday as part of Farmington's bicentennial observance. The flag pole is up and painted and Friday afternoon members of the crew installed the pulley and rope. A bulldozer was attempting to level the ground around the flag pole, men were painting, others were working on the fire station. The public is invited to the ceremonies. (Yeaton Photo)

The Lewiston Daily Sun

Lewiston-Auburn, Maine   Saturday, July 3, 1976   15

Picture Courtesy Lewiston Daily Sun
Dedication of New Municipal Building and Fire Station
Lower Main Street
July 3, 1976

Above Courtesy Town of Farmington 1977
Top-Bot, 1st Row L-R: Bob McCleery.
2nc Row: Office Staff: Fay Adams, Donna Tracy,
Beverly Beisaw. Nora Therrien.
3rd Row: Carroll Sinskie. Board of Selectmen
Ray Magno, Ron Wyman, John Ranger,
Larry Stofan, Lewis Fitch.

On LT: J. Bauer Small. On RT: Harvey Smith.
FFD Reps New Station Study Committee

Dedicated

for

Outstanding

Public

Service

This municipal report is dedicated to J. BAUER SMALL in recognition of his service to the Farmington Fire Department. Starting as a rookie in 1936, becoming a member in 1940, and serving as Fire Chief since 1951, retiring in Dec. 1969.

IN MEMORIAM

HARRISON W. BRAGDON

On Saturday evening December 18, 1976, the Company paid their respects to 1st Assistant Fire Chief Harrison "Hap" W. Bragdon who passed away on Wednesday.

The firemen met at 7:00 P.M. in the new Municipal Building Fire Station and marched to the Hawthorne Funeral Home where they paid tribute to a friend and Brother Fireman. There were 7 members in dress uniform who took part in this special evening visitation.

Sunday afternoon the Company went to the funeral of Assistant Chief Harrison W. Bragdon. There were 30 firemen in full dress uniform who paid their final tribute and respects to their departed Brother. At the conclusion of the service, Assistant Chief Bragdon's body was carried by pall bearers Fire Chief Forest Allen, Assistant Chief George Hobbs, Assistant Chief Robert McCleery, Fireman Harvey Smith, Fireman Herbert "Pete" Durrell, and Fireman Cyrus "Cy" Decker between an honor escort of firemen to Engine #5 which was used to convey the body to Fairview Cemetery.

"Hap" Bragdon joined the Farmington Fire Department in 1947. He was named Assistant Fire Chief in 1966 and promoted to 1st Assistant Chief in 1969. "Hap" was a member of the committee that drew up specifications for the Mack, Engine #5. He was very active on the Municipal Building Committee which included the new Fire Station. The new training program which we have all benefitted from was brought about by "Hap's" hard work and determination to make our Department one to be proud of. He will be long remembered for his service and devotion to the Farmington Fire Department.

A Tribute To

FOREST L. ALLEN

for Outstanding Public Service

Fire Chief Forest L. Allen has completed some 30 years of dedicated service as a fireman in the Town of Farmington. Chief Allen joined Farmington Fire Co. #1 in 1943. During these past years, he has seen many changes in fire fighting techniques and fire equipment. Many of the latest methods and apparatus have become a part of the Department during his years as an assistant chief in the 1960's and as chief since December 1969.

Chief Allen has given support to the Franklin County Firemens Association and helped make it an important organization of our area. He has, along with former Asst. Chief Bragdon, supported and initiated an extensive fire training program for members of the Fire Department.

During his years as Chief, the Department has obtained a new fire engine, moved to a new fire station and continued to give the residents of the Town a Fire Department second to none in service and dedication. The Town has also seen the merger of Farmington Falls and Farmington Fire Companies into a Municipal Fire Department.

Chief Allen is a member of several fraternal and civic organizations of this area, the Maine Federation of Firefighters and operates his own electric service business.

As Chief Allen enters his retirement, may we the citizens of Farmington take a moment to reflect upon his dedicated service and simply say, thank you for a job well done.

Dedication for Public Service

MRS. VERTIE ALLEN

The 1982 Farmington Town Report is dedicated to Mrs. Vertie Allen for her 13 years as Fire Department Dispatcher and for her faithful service to the citizens of Farmington in attending to their needs for emergency assistance.

Town Annual Reports Top-Bot, L-R: 1969, 1976, 1977, 1982

Robert "Apple" Oliver Sr.

In Recognition
of
49 Years
With The
Farmington Fire Dept.

IN RECOGNITION

We take this opportunity to recognize Norman E. Collins for his Forty six years of service as a member, and later Assistant Fire Chief of the Farmington Falls Fire Unit.

Norman was a Charter member of the former Falls Fire Department when organized in October of 1947. Later, that establishment became affiliated with the Farmington Fire Department in 1975 and he was named Assistance Chief. He has always been active in the annual Falls Firemen's Field Day and Famous Chicken Barbeque event. We trust he will continue these interests. Chief Collins retired March 1, 1993.

Thank you Norman Collins for your very long term community service and dedication.

IN DEDICATION

The Board of Selectmen would like to dedicate the 1996 Annual Town Report to Vertie and Forest Allen for over 40 years of service to the Town of Farmington. Mr. Allen joined the Farmington Fire Department in 1943, becoming Chief in 1969. That same year, Mrs. Allen began dispatching for the Fire Department until the Sheriff's Department took over those duties. Mrs. Allen continued to volunteer until 1991 giving out burn permits. The Allen's have the thanks and appreciation of the citizens of Farmington.

DEDICATION

CHIEF ROBERT "BOB" McCLEERY
(Photo by Luce Studios)

The Board of Selectmen has chosen to dedicate the 1999 Annual Farmington Town Report to Fire Chief Robert L. McCleery. He became a member of the Farmington Fire Department in 1960 and has served as Chief since April 1977. He will be retiring in June 2000.

Under his leadership, the Fire Department has expanded greatly with the mandated changes being a challenge. Countless hours were necessary to keep abreast of the regulations and to implement them. The Fire Department is one of several organizations that Chief McCleery has served faithfully and many times as an officer.

At the Maine Fire Chief's Association annual meeting in October 1998, Bob was honored as "Chief of the Year". This honor reflects the high quality of the Chief and members of Farmington Fire Department.

Our sincere appreciation to Chief Robert L. McCleery for his commitment to serve our community in this capacity.

Town Annual Reports Top-Bot, L-R: 1985, 1994, 1996, 1999

Robert L. McCleery 241

CHIEF J. BAUER SMALL
1951 - 1969

Annual Reports

Report of Chief of Fire Department

To the Inhabitants of the Town of Farmington:

To my knowledge there has never been a report of the Fire Department in the Annual Town Report. This has been due to the fact that the Fire Company is maintained by the Farmington Village Corporation. By maintenance, I mean they provide the station, equip us with boots, coats, etc. and pay us a set sum each year for all fires inside the Corporation limits. As the town proper contributes its share toward the operation of the Department I feel it only right to inform them of our activities.

During the year we answered a total of 84 calls. These consisted of 18 calls inside the Corporation, 4 in the Suburban Water District, 60 within the town limits and 2 in New Vineyard. The following is a breakdown of the town fires:

Chimney	31	Grass	2
Building	17	Short circuit	1
Car	1	Dump calls	5
Brush	2	False alarm	1

We have been most fortunate in not having any large scale fires. The Butler property on Pleasant Street was our major loss of the year, at an estimated $18,000. As you can see we are still busy on chimney fires, but these are minor compared to years back due to oil installations and, I believe, people today seem to be more fire conscious. Grass fires in the Spring were a major headache, but now these are at a minimum.

As you will no doubt notice, there is an article in the warrant to raise the sum of $7,000, approved by the Budget Committee, to put with the $2,000 raised last year for the purchase of a new tank truck. This tank truck will be of a 1,000 gal. capacity with a 500 gal. per minute front end pump. The purchase of this truck has two prime factors in its favor. At the present time we maintain one pump capable of pumping at areas outside the hydrant district. This truck will give us another pump with a total capacity of four lines in conjunction with its tank load. The second factor is that we will place our present tanker at Farmington Falls. They have an excellent volunteer department there which is wholly self-maintained. The tanker will be maintained by them but will still be the property of the town with the understanding if we need its services in any part of the town it will be sent on call. It will also provide a protection for property there.

I would like to stress something that I believe to be of the utmost importance to those people in the outlying areas. The Osborne fire is still fresh in our minds. We never could have saved what we did without their having a farm pond. This, in itself, should be an incentive to those who have the facilities, to have one for their own protection. Also, incidentally there is nothing that makes a department feel more helpless than stand around with all kinds of equipment and no chance to use it.

This report is for the Town of Farmington only.

Respectfully submitted,

J. BAUER SMALL, Chief.

1954 First Fire Chief's Report to be Included in Town Annual Report

Report of Chief of Fire Department

To the Inhabitants of the Town of Farmington:

I hereby present my report of the activities of the Fire Department during the year 1955.

We answered a total of 44 calls. These consisted of 8 inside the Corporation, 2 in the Suburban Water District, 31 within the town limits, 1 in New Vineyard, and 1 in Chesterville. The following is a breakdown of the town fires:

Chimney	18
Car	1
Building	5
Oil Burner	1
Dump	5
Brush	1

All in all, I can find no year in the back records where we have had such a small fire loss. Last year we thought rather exceptional with 84 calls while this year the total was a little more than half. Our only serious fire was the loss of the Harold Hardy buildings on Perham Hill. Of course, we are still blessed with a large percentage of chimney fires, 90% due to lack of cleaning.

We received our new pumper tanker in July. We put it in service immediately and transferred the old tanker to Farmington Falls as planned. The new truck has given us exceptional service so far and the company is very well pleased with it. The tanker at the Falls has extinguished several fires which could have resulted in serious losses. We have purchased some hose, 4 lights, a battery charger, a hose washer, and several small additions to the equipment.

On speaking of our new tanker, I would mention that following its arrival we planned an Open House, to show the townspeople this new piece of equipment, as well as our other equipment. On the day of the Open House we had a mere handful of visitors. This small showing was very disappointing, as we know you are interested from past experiences. We are justly proud of our station and trucks, and anyone is welcome to inspect them at any time.

The town appropriates us the sum of $2,000 per year. This sum represents a very small expense to the town for services rendered. I would like to suggest that each year we have an unexpended balance in this appropriation that it be automatically transferred to an equipment fund. This wouldn't be any large sum each year, but, if we had some kind of a reserve, and had a serious hose loss or some other piece of equipment, we could replace it and possibly avoid an overdraft.

We do not have a house to house yearly fire inspection as we haven't the available men that have the time to give for such services. If you have any fire problem, please get in touch with us as we would rather come to your house in an advisory capacity rather than an alarm at three in the morning at twenty below zero.

This report is for the Town of Farmington only.

Respectfully submitted,

J. BAUER SMALL, Chief.

1955

Report of Chief of Fire Department

During the year 1956 the department answered a total of 60 fires. The greater part of these were minor fires, 23 of these being chimneys. We had two rather bad ones, the Kenneth Ladd house on the Strong Road and the Perkins residence on the Holley Road. One fire I believe set a record when we were called to the Tom Moore farm on the Temple road for a grass fire January 28th.

The department is in excellent condition. The new tanker is working out very satisfactory and the Falls Department is doing a great job with the tanker we sent down there. We are experiencing some trouble about hearing our alarm. We should have it remedied in about a month. We find it hard to compete with television.

On the whole we have experienced a very quiet year. I would like to advise once more that we will gladly make a building inspection for any person desiring one.

Respectfully submitted,

J. BAUER SMALL, Chief.

1956

Report of Chief of Fire Department

To the Town of Farmington:

I hereby submit my report as Fire Chief for the year 1957.
The following report is for the town only.
We have answered a total of 48 calls as follows:

Town:
    Chimneys 11
    Buildings 4
    Oil Burners 2
    Spilled gas 1
    Tree afire 1
    Tie culvert 1
    Milk tank 1
    Sawdust pile 1
    Grass 9
    Dump 2
    Woods 3
    Auto 2

Suburban Water District:
    Chimney 2
    Grass 2
    Smoke scare 1

New Vineyard:
    Chimney 1
    Building 1

New Sharon:
    Building 1
    False 1

Chesterville:
    Building 1

The department has experienced a rather quiet year. We have had no serious fires except for the Wesley Mitchell and the Judkins fires.

The equipment is in good order and we are certainly making good use of the new tanker. Also, by placing the other tanker at the Falls they have used it to a great advantage and have saved the town quite a lot of money by our not having to answer to minor fires in that area.

Our article this year asks for an additional $500.00. The main purpose of this increase is for new hose. Also, like other departments our expenses have increased accordingly.

I would strongly recommend to the town that at some meeting in the near future that an equipment fund be set up for the fire department in the sum of $2,000 a year. Our Maxim pumper is now 25 years old and will have to be replaced soon. It isn't that these trucks wear out, but they become obsolete and parts aren't available. A new one similar to the old now costs from $12,000 to $15,000. Consequently, with some sort of an initial beginning it would soften the blow of the full amount, which could happen at any time.

Respectfully submitted,

J. BAUER SMALL, Chief.

1957

Report of Chief of Fire Department

To the Town of Farmington:

I hereby submit my report as Fire Chief of the Town of Farmington for the year 1958. This report is for the town only, as I also submit one to the Farmington Village Corporation for fires in its area.

We have answered a total of 57 fires, which in comparison to other years, is rather light. 11 of these were in the Corporation, 2 in the Town of Industry, and one in the Town of Chesterville. The following is a breakdown of the town fires.

Buildings	7
Chimneys	17
Grass	4
Auto	5
False	2
Oil Burner	4
Dump	2
Air blower	1

As you can see, we are still long on the chimney fires. The strange thing is, that these chimneys are the same ones, practically, that we visit each and every year. If these chimneys would receive the proper care, our calls would be cut in two.

There is an article in the warrant to appropriate $2,500.00 for an equipment fund. I have mentioned in previous reports, that our Maxim pumper is beyond the age limit allowed by the insurance underwriters. They go by a ruling that, although the piece of apparatus may be mechanically sound, after twenty years it is considered obsolete. Our Maxim was purchased in 1933, so it is well over the time limit. This amount of money which we are asking, does not come anywhere near the purchase price of a new pumper, which is now in the vicinity of $14,000, but will give us something to start on.

I think the town should establish a permanent equipment fund, such as the Highway Dept. Even a small sum appropriated each and every year would take care of our equipment needs without asking for a large sum at one time.

The Department is in good shape. The roster is full and we have a waiting list. We haven't added any new equipment this year, but we hope to at least radio equip the tanker sometime this Summer.

Respectfully submitted,

J. BAUER SMALL, Fire Chief

1958

Report of Chief of Fire Department

The following is a report of the Farmington Fire Dept. and does not include fires inside the Farmington Village Corporation.

- 11 Chimneys
- 2 Faulty Wiring
- 4 Building Fires
- 2 Woods Fires
- 2 Hot fat burning on Stove
- 1 Town Dump
- 1 Flooded Oil Burner
- 1 False
- 1 Grass
- 1 Smoldering Cigarette
- 1 Call to Chesterville
- 1 Call to New Sharon

This past year has been an exceptional one for fire calls, especially with the length and rigorous cold of the winter.

There are two articles in the warrant this year relative to The Farmington Fire Department, one to transfer the Department from the Farmington Village Corporation to the Town of Farmington, and the other for the support of the same.

The Farmington Village Corporation was first organized primarily for fire protection, and, at that time, the fire department was only capable of giving service to those buildings inside its limits. However, with the equipment maintained today, the department services the entire town. The Farmington Village Corporation has equipped the personnel of the department, furnished quarters, and paid a sum of money for protection within its limits ever since it was inaugurated through taxation of those individuals owning property there. Inasmuch as the department has outgrown its original service, I think it only fair that the entire town assume the total cost of maintaining it.

I would like to mention in this report that the Fire Department and the Water Department were thoroughly inspected last summer by the Board of Underwriters, who determine our insurance rating, and we were granted a rating that few volunteer departments are fortunate to acquire. This rating was contingent on our replacing our Maxim pumper, for which we are asking another sum of money this year.

Very respectfully submitted,

J. BAUER SMALL, Fire Chief

1959

Fire Chief — J. BAUER SMALL

Report of Fire Chief

To the Inhabitants of the Town of Farmington:

It gives me great pleasure to submit my tenth annual report to the Town as Fire Chief.

This past year was a busy one with 73 calls answered, not counting numerous inquiries and investigations of dubious fire conditions. Chimneys and flooded burners still head the list, but we had three serious fires during the winter. Any fire is serious, no matter in what proportion, but these particular three nearly involved fatalities. We read every day of fatal fires, sometimes wiping out whole families, but it doesn't sink in until it happens to someone you know. These fires, namely Ferrari, Newcomb and Petrie, involved young people who were active and alert, and managed to get out of the building. Had these persons been of an advanced age I would hate to have predicted the outcome. Consequently, I urge all possible precautions.

The department roster is full. We had several resignations and one member went into the armed services, but their vacancies have been filled. All equipment is in excellent condition. I hesitate even mentioning a new truck after the performance of our 1931 Maxim pumper at the Bailey fire but, as you can see, it has passed its 30th birthday and the Insurance Underwriters still consider a truck obsolete after twenty years. At the present time we have some over $15,000 in our reserve equipment fund which is nearly the amount needed for a replacement.

We are participating this year in a state-sponsored fire school which will be most beneficial, especially to the newer members.

The Town, this past year, has installed a radio base station at the Community Building for communication with town equipment. I feel we should, this year, install a relay at the fire station and a radio in the tanker. I don't think it necessary to equip the other trucks as the tanker is always at the scene.

I am reverting, once more, to our alarm system. As I have explained in past reports we are still having trouble hearing our present alarm and I spoke of having radio receivers in each member's house. With a relay unit at the fire station from the base at the Community Building these units could be activated thus alerting all members. The Town, in the past, has appropriated a sum of money toward the equipment fund and I would suggest that this year an amount be appropriated but used toward the purchase of these units. This particular type of alarm system is being installed all over the country and has proven to be the answer to situations exactly like ours.

Another sobering financial problem is that, eventually, we will be forced to maintain permanent men at the fire station. This is not an immediate necessity but, as the Town grows and the Underwriters demand more protection, it certainly will be a requirement. At the present time, however, we enjoy the highest insurance rating possible under our voluntary organization. But, like all the municipal departments, there is a constant demand for more and varied services. A rather extreme case in point was a person who called on the fire phone and asked how they would go about to dispose of a dog. My first thought I left unsaid but I told him to contact a veterinarian. I further asked him how he happened to call the fire department for such information and he told me that where he came from they were instructed to call the fire department in case of any emergency.

The department is honored to have this Town Report dedicated to them. We will certainly strive to maintain our record and fulfill the responsibility you have placed in us.

Respectfully submitted,

J. BAUER SMALL,
Fire Chief

1961

Pictured L-R: Cy Decker, Richard Russell, Jack Bell, Howard White

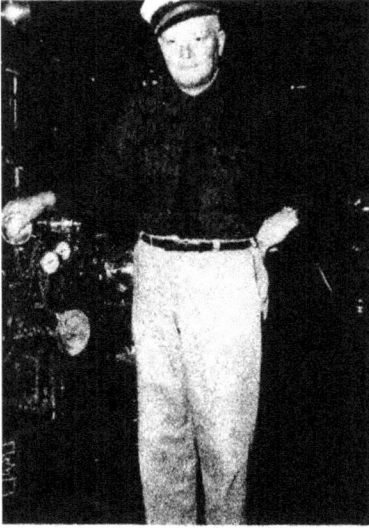

Report of
Fire Chief

To the Inhabitants of the Town of Farmington:

It doesn't seem possible, the years going as fast as they do, that I present my twelfth report to you as Fire Chief.

We answered 62 calls this past year and, as usual, chimney fires predominate. It has always been a mystery to me how so many chimneys can get afire after having been cleaned only a few weeks before. It is rather amusing to have a husband say the chimney has been cleaned only a few months previously and then the wife, inadvertently, telling us later that it hadn't been cleaned for three years. In a case of this type we are more apt to believe the woman.

There was only one serious fire, that of Hubert Knowlton on Perham Street. It certainly is a good feeling to see these places rebuilt, even better than before because, when we lose one in the country, and we do lose them occasionally, it is hardly ever rebuilt.

We had a State-sponsored Fire School during the Winter on hose and ladders which was well attended. These schools are very important to volunteer departments. They are set up by the State at no charge to the Town.

Our radio alerting system is now complete. We have 22 units, one in every fireman's home, and the three chiefs have red fire phones. It has proved to be most valuable, and will be more so as we go along. At present the entire system is located in my bedroom. This has worked out well as far as alerting the the company, but there is one draw back. Our chief night dispatcher is apt to go back to bed and asleep. I might say, in deference to her loyal service, she is doing better. Eventually, as the Town grows and we are obliged to maintain some permanent men, this will be transferred to the Fire Station.

The company is complete, and we have a waiting list. We are most fortunate in having such a group of able and dedicated men in our department.

As you will notice in the warrant we are asking for another $1,500 for our equipment fund. We are starting to lay groundwork to purchase a new pumper to replace our 1931 Maxim. Rather than turn in this piece of equipment, which operates perfectly, but the age is against it, I think we should put a little more money in it and convert it into a utility truck with a larger tank. This would give us three pumpers and four tanks, counting one tanker at Farmington Falls.

I might say, in closing, that if there are any suggestions for improvements that anyone thinks would be for the good of the department we would gladly welcome them.

Respectfully submitted,

J. BAUER SMALL

1963

Report
of
Fire Chief

J. BAUER SMALL

It is with pleasure I submit my sixteenth annual report to the citizens of Farmington.

We lazed along this year with only 59 calls compared to 96 last year, with none of these of major proportion. The loss of the out barn at Voter farm and the Farmington Supply Company were the only two to run to any extensive loss.

The dump seems to require our attention about the same number of times each year, along with chimneys and oil burners. For same reason, we had only one grass fire all year. I have seen the time we had five going at one time, which had a tendency to spread us a little thin. I hope this trend continues.

Car fires seem to be increasing, not only with us, but statewide. It is rumored that sometimes the payments are a bit too large or too close together. We haven't, knowingly, run into this sort of financial easement.

We received delivery of our new truck in February and are very much pleased with it. We haven't had a fire, rap on wood, where we could really see what it will do, but, so far, it has performed very efficiently. I believe we have been instrumental in selling quite a few of these trucks, as numerous departments have been to see it and have ordered near duplicates.

The company stands at full complement of 25 and we have one new rookie. We are at present having a First Aid course, through the courtesy of the Red Cross. We felt this to be essential as we are being called upon more and more for this type of service.

We had quite a "pins and needles" situation this Fall with the water shortage. We had eight tankers available that would have handled, by shuttling, any normal fire. If we had had a real smoker, we would have had to pump out of hydrants and use all the water in the transmission lines, thereby putting everyone on an equal basis with no water, plus a few capsized water tanks. I'm more than glad to report this situation is now remedied and my last report was that the reservoir was within a few feet of the top.

We have an article in the warrant pertaining to adding another $1,000.00 to our Reserve Account. I feel this is most essential as, when the time comes to purchase a new piece of apparatus, we have something to start with and not have to raise the total amount in one lump sum. Incidentally, it gets by the Budget Committee a bit easier.

I am still, and always will, continue to try to get something on a permanent basis as far as fire personnel is concerned. Granted, with the number of fires this year, it would seem to be a poor time to mention it. We, at the present time, have a good set-up, with the fire phone and the alarm button at the store, where it is attended all day, and between Anna, Frank Gagne, the other chiefs and myself is attended nights. This situation can change, and we should be giving it some thought.

In closing, I would like to bring to your attention that we operate on an approximate budget of $8,000.00. As far as I know, we are the only fire department in the country that pays rent for their station. This rent is paid to the Farmington Village Corporation and comes out of our appropriation. In contrast, the City of Rockland with a population of double our size, appropriates over $100,000.00.

Respectfully submitted,

J. BAUER SMALL

Fire Chief

1966

Report of Fire Chief

J. BAUER SMALL, Fire Chief

I hereby submit my annual report of the Farmington Fire Department to the inhabitants of the Town of Farmington.

I have been mentioning in my report for some years that something on a permanent basis be established. We will be working the rest of the year to bring about a change which, I hope, will be satisfactory to all. We have grown through the years; the fire telephone must be attended at all times and there are numerous duties that need to be done by a permanent chief that are impossible for me to do in my limited time. Also, as the University of Maine complex continues to grow, the insurance underwriters, to keep our excellent insurance rating, may insist on such a change. I don't intend to scare everybody by this "permanent" phrase, as it will probably entail no more than three or possibly four men. The department will still be a call department.

We have been most fortunate over the years in the cost of maintaining the fire department. As you can see in the appropriations we are asking $8,500 this year of which $1,200 will be earmarked for rent to the Farmington Village Corporation. Skowhegan's appropriation is $35,000, Dover-Foxcroft, smaller than Farmington, $13,000, and Rockland, some over 10,000 population, receives over $100,000. I realize this a most inappropriate time to even mention money but it has come, and by the way, I'll miss my $300.00 per year for being chief.

We have had a normal year with 69 calls. Chimney and grass fires were tied with 13 each. The others were of the usual — cars, oil burners, etc. We had no serious ones.

The department is fully manned. We have had one retirement which was filled by a rookie. We have had a State sponsored fire school this winter on assimilated fires in various buildings and are to have a pump school in the spring. Our equipment is in fine shape but we will probably have to replace some hose this summer.

I do want to thank all you people for your support and going along with allowing us to purchase all the improvements we have made over the last twenty years. There has never been a time we were refused except once, which I think bears telling.

I appeared before the Budget Committee in hopes of replacing our old tanker. Every member who was on the Board, and practically had a hydrant in their front yard, was in favor of it, but those who lived out in Greater Suburbia, to whom it would be more advantageous, turned it down. One, in particular was most vehement. That summer we were called to a house fire in Chesterville and were fortunate to extinguish it with what water we were carrying. The above mentioned person was on the sidelines. Several days later he came into the store and asked me if I was intending to try for a tank truck again that year. I informed him that I most certainly would do so. He said, "Go ahead. Now that I have seen the d - - - - thing work, I'm all for it".

Respectfully submitted,

J. BAUER SMALL
Fire Chief

Looking down upper Broadway in Farmington. February 1969.

1968

Fire Department: l. to r.: Philip Folger, retiring member, Forrest Allen, newly appointed Chief and J. Bauer Small, retiring Chief.

Report of the Fire Chief

To the Inhabitants of The Town of Farmington:

I hereby submit my last report as Fire Chief. It is rather hard to give up associations made during a 35 year stint in the department but, either I am slowing up or the thoughts of leaving a warm bed at all hours and going into sub-zero weather may be of some influence. As you know, Forrest Allen has been elected to take my place. Forrest has had many years experience in the department and I am sure he will do a good job.

As this is the last year you will hear from me on this subject, which I have included in previous reports, we have got to appropriate money for the fire department. In the first place we have to be thinking of something along a permanent men proposition and, secondly, a new fire station. Now, the first one will, no doubt, be one of the requisites demanded by the insurance underwriters. You might as well face this, as, if we don't follow their recommendations, they will change our rating and you will pay more for insurance than it would actually cost for permanent men. Personally, I think in a town this size that an entirely volunteer fire department is fast waning. We will always, no doubt, have the greater part of the department made up of call men, but your key men will be on a payroll.

Now, to the second part in regard to a new station. Our present one is rapidly becoming too small. We have three trucks on the floor and one standby in the basement. If we are required to have an aerial ladder we have no place to put it. This has been talked over for as long as Millard Quimby was here so it is nothing new. It is just reaching a point. Tentatively, it would have five bays, facilities for permanent men, bedrooms, kitchen, etc., space for Town offices and also room for the police department, instead of the closet they now call headquarters. I don't know if you are aware that we pay rent to the Farmington Village Corporation for our present station. It may vary from year to year as we pay 60% of all the expenses of the building. A round figure of $1,000 per year is average, so why not put this money into something the Town owns.

In reading this over and noting I have said 'We have got to appropriate money for the fire department' I have in mind the protection of all this taxable property which pays for you know what. Several big losses could result in higher taxes.

I do want to thank the Town for its cooperation during my stay in office. They have been more than generous in supplying equipment we needed, and we do have an excellent department.

Always, when I suggest money, I have to report a light year. We had the usual chimney, oil burners, cars, etc., but no major fires.

Respectfully submitted,
J. BAUER SMALL

1969
Chief Bauer Small Last Annual Report

Chief Forest L. Allen
1969-1977

Report of
the Fire Chief

To the Inhabitants of the Town of Farmington:

The following is the annual report of the Fire Department for the year 1972. The Department has answered 69 local calls and 5 out of town calls.

We have held 12 regular meetings and 2 special meetings. We have had a school of instruction conducted by the state in the use of hose and pumps.

Last year in my report I brought to the attention of the townspeople the fact that the Fire Department felt that serious consideration should be given to building a new fire station.

The building now used as a fire station is owned by the Farmington Village Corporation. This means that the Fire Departments pays $1200 a year plus 60% of the maintenance.

At present the town holds a Class C insurance rating. If this rating is lost due to lack of proper equipment and housing facilities, it would mean that the property owners insurance rates would increase 10-20% and in some cases would be doubled.

We as members of the Fire Department, feel that all citizens of Farmington should bear in mind that this is a college town. Each year the college has a larger enrollment and new and larger buildings. We feel our equipment is not adequate for proper protection of the town. Before adequate equipment can be obtained, we need larger housing facilities.

The Department hopes the voters will seriously consider the building of a new fire station.

Respectfully submitted,

FOREST L. ALLEN

1972
(Town Annual Reports for 1970, 1971, 1974 not found)

REPORT OF THE
FIRE CHIEF

To the Inhabitants of the Town of Farmington:

The year 1973 was a relatively quiet one for the Fire Department. The Department answered a total of 79 calls; 69 of these were local calls and 10 out of town calls. We had no very serious calls, but we did notice an increase in the number of chimney fires.

The Department held 12 regular meetings and a few specials.

In 1973 the Department purchased 500 ft. of 1½" hose, two 2½" nozzles and one 1½" nozzle.

The members of the Fire Department wish to express their thanks to the people in town who have shown an interest in eventually having a new Fire Station and equipment for the Town of Farmington.

Respectfully submitted,

FOREST L. ALLEN
Fire Chief

1973

Report of the Fire Chief

To the Inhabitants of the Town of Farmington:

The Fire Department answered a total of 73 calls in 1975; a decrease from the previous year. The total calls consist of the following number and types: 10 car and truck fires, 13 chimney fires, 7 house fires, 5 business fires, 5 grass fires, 3 calls to the town dump, 4 gas spills, 5 false alarms, 4 calls to UMF, 2 flooded oil burners, 4 car accidents, 2 power lines down, 1 television set and 6 outside calls. The outside calls were 4 to Industry and 2 to Chesterville.

We had one major fire which resulted in the destruction of the Drummond Hall Block.

The Department held 12 regular meetings and 5 special meetings; also 4 forest fire schools, 4 pump schools and 14 fire training schools.

The Department purchased a new Mack truck equipped with a 1000 gallon a minute pump, 4 MSA Airpacs and 1000 feet of 2½ inch hose.

In 1975 the Farmington Falls Fire Co. became part of the Farmington Fire Department.

We are looking forward to moving into the new fire station.

Respectfully submitted,
FOREST L. ALLEN
Fire Chief

Farmington Fire Department.

1975

Report of the Fire Chief

To the Inhabitants of the Town of Farmington:

The Farmington Fire Department moved to the new Fire Station located on Lower Main Street on December 10th & 11th, 1976. The firemen are all very appreciative of the new station.

The Fire Department answered a total of 74 calls in 1976. The total calls were as follows: 9 car and truck fires, 3 fires in the business district, 4 accidents, 16 chimney fires, 3 calls to UMF, 9 house fires (2 fatalities), 6 false alarms, 2 flooded oil burners, 4 grass fires, 2 woods fires, 1 call to FCM Hospital, 2 barns and sheds, 6 outside fires, and 7 miscellaneous calls. The outside calls were 1 to Industry, 1 to New Sharon, 1 to Wilton; in each of these a house was destroyed. There were also 2 calls to Temple and 1 to Chesterville.

The Department held 12 regular meetings and 4 special meetings; also 21 training schools; 12 of these held at Farmington and 9 at Farmington Falls. There were 13 work meetings called in the process of moving to the new station.

In 1976 the Department purchased uniforms, badges and red lights for all firemen, one 1½" and one 2½" Turbojet nozzles, 5 radio receiving sets for Farmington Falls, 14 helmets and safety goggles, 2 MSA Airpacs and spare tanks, 1000 feet of 1½" hose and 500 feet of 2½" hose, 2 Floodlights for the portable generator, boots for Farmington Falls. The Farmington Falls Department is now supplied with all necessary equipment. Also ordered is 1600 feet of 4" supply line which should be arriving soon.

Work has begun on finishing the upstairs for the Department.

The people of Farmington should be made aware of the fact that there is a need for an aerial ladder truck for the Fire Department.

Respectfully submitted,

FOREST L. ALLEN
Fire Chief

Fighting West Farmington Fire

Preparing to wash down spilled gasoline at accident scene

1976
Chief Forest Allen Last Annual Report

Chief Robert L. McCleery
1977-2000

Report of the Fire Chief

To the Inhabitants of the Town of Farmington:

I hereby submit my first annual report of the Farmington Fire Department since assuming the office on April 1, 1977.

During the past year the Fire Department answered 135 calls. Included in these were out of town calls (Chesterville, Industry, New Vineyard, and Wed), accidents, cars, false, grass and woods, oil burners, commercial (Maine Dowel, Macomber Mill), house and miscellaneous calls (overheated stoves, electrical outlets, electric motors, downed power lines, transformers, sawdust piles, wood piles and a paper dispenser at Mt. Blue High School).

There were several chimney fires due to improper use of wood burning stoves. I cannot urge the citizens of Farmington enough in the proper installation, care and maintenance of their chimneys and wood burning stoves. The Fire Department has material available on the proper installation of wood stoves. I strongly recommend the use of some type of a warning device, either heat or smoke alarm, but good common sense is still the best method of fire prevention.

The important need of both the Fire and Police Departments is for 24 hour dispatching service in the Municipal Building. At the present time all fire calls are dispatched from former Chief Allen's home.

Under the direction of our Training Officer, Deputy Chief Glenwood Farmer, various training sessions were held. Among the subjects covered were C.P.R. and First Aid courses, evacuation drills, different types of pumping and hose lays, self-contained breathing equipment and arson detection courses.

In the past year the I.S.O. (Insurance Services Office) conducted a rating survey of the Fire and Water Department. The Town has not received the results of the survey, but a copy of the recommendations for upgrading and improving the Fire Department has been received. This includes the need for a full-time Chief and sufficient company officers, a new 1,000 G.P.M. pumper and an Aerial Ladder Truck of modern design and a pumper tanker for the Falls Substation.

The Town is not far away from some permanent Firemen as it is continuously harder to summons day time help for fires. Also more demands are being made of the Fire Department in regard to chimney inspections and building permits from the Fire Marshall's Office for Day Care Centers and Foster Homes.

In September a Squad Truck (Rescue) was added to the Fire Department fleet. This vehicle, with its trained personnel should add much to the efficiency of the department.

We thank Hawthorne Ambulance Service for their response to all our fire calls and appreciate their support in our training programs.

I would like to take this opportunity to thank the citizens of Farmington for their cooperation and my officers and men for their support.

Respectfully submitted,
Robert L. McCleery
Fire Chief

1977

As Included in the Fire Chief's Report 1977

FIRE DEPARTMENT

1st Row; *Melvin Bard, Gerold Cookson, Mike Knudtson, Thomas Cassidy, Lt. Clinton Blaisdell, Kenneth "Spider" Durrell, Charles Grant, Ivan Howard Jr., Terry Bell*

2nd Row; *Harvey Smith, Asst. Chief Clyde Ross, Chief Robert McCleary, Deputy Chief Glenwood Farmer, Asst. Chief Richard Russell, Asst. Chief Norman Collins, Capt. John Bell, Bradford Moore*

3rd Row; *Timothy Hardy, Lt. Clyde Meader, Douglass Oliver, Roy Hazzard, Morrill Collins, Roger Ladd, Howard White, Carroll Corbin*

4th Row; *Junior Turner, Robert Oliver, David Ferrari, Mahlon Moore, Herbert "Pete" Durrell*

Report of the Fire Chief

To the inhabitants of the Town of Farmington, I hereby submit my annual report of the Farmington Fire Department for the year 1978.

This year the Fire Department answered 160 calls, included in these were out-of-town calls, highway vehicle accidents (cars, trucks, cycles), grass and woods, houses, overheated stoves, electrical outlets, warehouses, miscellaneous and training. Chimney fires continue to receive the largest number of calls due to improper wood burning and sooty chimneys. Along with these calls, there was an increasing number of requests to inspect chimneys, buildings and homes.

An important need for both the Fire and Police Departments is full, 24 hour dispatch service from the dispatching office in the Municipal Building. There is still a need for some full time firemen (maybe 3 men at the station during the day) to answer emergency calls.

The new Aerial Ladder Truck should be ready for delivery by late March or April. Looking ahead, plans should be made and perhaps drawn up for the purchase of a new 1000 G.P.M. Pumper Truck to replace Engine #3.

This year we started to implement the N.F.P.A. Fire Fighter I, II, III criteria in our Training Programs. Some of the subjects covered and to be expanded upon are Fire Tactics, Protective Breathing Apparatus, Pumps, Driver Safety and the Handling of Hazardous Materials.

In closing I want to thank the citizens of Farmington for their cooperation and support of the Fire Department's Activities. Thanks goes to the Officers and Firemen for their dedicated efforts.

Respectfully,

Robert L. McCleery,
Fire Chief

1978

REPORT OF THE FIRE CHIEF

To the inhabitants of the Town of Farmington, I hereby submit my annual report of the Farmington Fire Department for the year 1979.

This year was a relatively quiet one with fifty-seven less calls than last year, for a total of one hundred and three. Chimney fires continue to makeup the largest number of calls as more homes are being heated with wood. I must continue to remind the citizens to carefully check their wood stove installations, and chimneys for soot build up. We are receiving more requests for chimney inspections than ever before

Four homes (two mobile) were totally destroyed and several others received fire damage. Service calls, false alarms, accidents, vehicles, oil burners, grass and brush, mutual aid, and an air craft crash made up the remaining fire calls. In February we had two fire calls simultaneously, one a mobile home on Titcomb Hill and the other an apartment house on Perkins Street, which made a busy evening. The winds on this night were twenty to thirty knots and the temperature about six degrees.

The need for twenty-four hour dispatching for both the Fire and Police Departments, possibly a "911" system, is becoming more apparent each year. In the not too distant future, there will be a need for some full-time firemen here at the Fire Station.

The new Aerial Ladder Truck has arrived and should be in-service before this book goes to press. Specifications on a new 1000 G.P.M. Pumper Truck should be put together during 1980.

The Department's training programs are well attended and progressing well. This year training with the Aerial Ladder will be the priority. As in the past years, we are receiving continued support from the State Fire Service, Department of Education and Cultural Services in Augusta. The top notch instruction gives our firefighters the latest information and methods on fire tactics. We have continued to drill and practice with Protective Breathing Apparatus, Pumping Units, and Fire Ground Tactics.

The Fire Prevention Program, used in the local elementary school system, was supported by the local insurance agencies. The Captain No-Burn program was very well received and more than three-hundred students participated.

In conclusion, I would like to thank the citizens for their continued support of the Fire Department, its programs, and being patient. We have a well trained, efficient and dedicated group of Firefighters, let's support their efforts.

Respectfully submitted,

Robert L. McCleery
Fire Chief

1979

Robert L. McCleery 267

1980 REPORT OF THE FIRE CHIEF

To the Inhabitants of the Town of Farmington, I hereby submit my annual report of the Farmington Fire Department for the year 1980:

This year saw a near record in fire calls established with chimney fires again contributing the greatest number. I must continue to remind the citizens to carefully check their stoves and chimneys. Also, to be careful in the disposal of ashes. There have been several fires caused by ashes placed in plastic pails, bags, or cardboard boxes and then dumped beside a building. Chimney inspections are again on the increase this year as more people are turning to wood and coal for heating purposes.

The dispatching service has been expanded to cover the towns of Chesterville, Industry and Temple. These along with New Sharon and Wilton give us close communication with our mutual aid towns.

A fire truck committee has been appointed to draw up a proposal for a new 1000 gallon per minute pumper. This should be ready for presentation at Town Meeting.

The training programs were well attended and broadened to cover a variety of fire-related incidents with emphasis placed on the new Aerial truck. We are continuing to receive support from the Department of Education and Cultural Services in fire training.

Through the combined efforts of the State Fire Marshal's Office, Farmington Police Department and the Farmington Fire Department personnel, several arson cases were solved the past year. This resulted in a conviction and a State's Prison sentence.

The Fire Prevention Program, Captain No-Burn, was again used in local elementary schools. This is sponsored jointly by local insurance agencies and the Fire Department.

In conclusion, I would like to thank the citizens for their continued support of the Fire Department and its programs. Through these efforts, we continue to have a well trained and efficient department.

Sincerely yours,
Robert L. McCleery
Fire Chief

1980

Assistant Chief S. Clyde Ross
1977 - 1991
Deputy Chief & Training Coordinator
1991 - Present
(Picture taken 1983)

TRAINING REPORT, FIRE DEPARTMENT

To the Town Manager, Selectmen and Citizens of Farmington:

Now that 1981 has come to an end, it is time to evaluate the
department's training activities for the past year. There were 21
Training Sessions held between Farmington and the Falls sub-station.
This was less than in 1980, but much was accomplished.

It is often difficult to convince members of the need to have a
continuous training program, one that each man should feel a part of.

To make this more real, we tried to train in smaller groups or by units
and engine companies. This gives each person more opportunity to be
actively involved in the training, not just standing around and observing.
Hands-on is more effective as a method to have men learn and become
familiar with fire fighting apparatus and equipment.

Training sessions were held where instruction was given in use and
maintenance of Protective Breathing Apparatus, C.P.R., Hydrant hookups
and Relay Pumping, Aerial and Ground Ladder uses, Engine Pump Operations,
and an Evacuation Drill at the University of Maine at Farmington with
Keegan Ambulance and College units. This was a cooperative effort and
has worked well to establish good public relations. Attendance has
averaged around 14 men per session.

In August Terry Bell attended the Fire Fighting Academy in Presque
Isle where he qualified as a Fire Fighter I and II. This is an honor
and a first for our department. Four men attended the Annual Fire School
at the University of Maine at Orono in June. This is a very interesting
session as there are top instructors and men from all over the state.
There is time here to share notes and compare department operations and
practices.

In July we hosted the first Franklin County Fire Attack School.
Several of our men were instructors for this well-attended session. We
gained much knowledge as well as recognition from other departments
when they realized the expertise of our men and the equipment we have
for fire attack. Farmington can be very proud of its men and equipment.

I would like to thank those men who have assisted in the training
program at Farmington and the Falls. Without the continued dedication
and interest we would be hard pressed to have an effective Training Program

Respectfully submitted,
S. Clyde Ross
Assistant Chief

1981 Training Report
Assistant Chief S.Clyde Ross

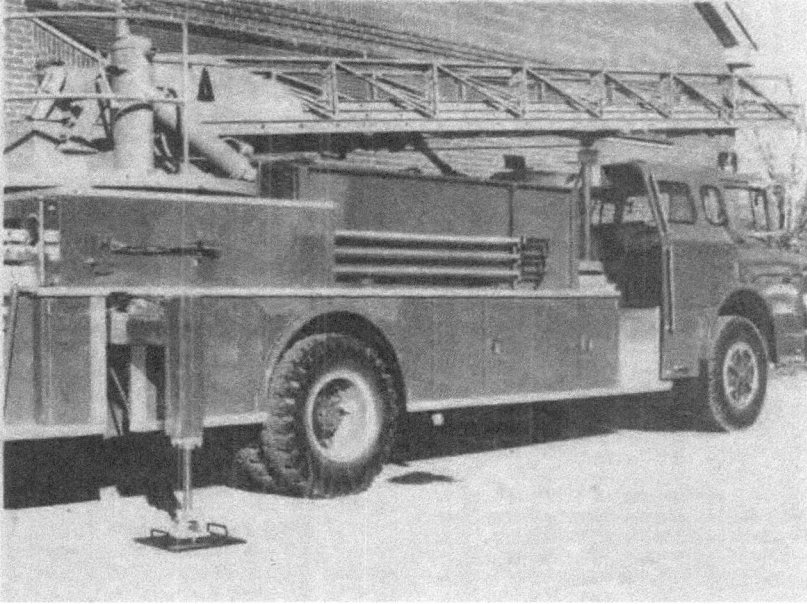

1979 Ford C - 900 Truck with used 75 foot ladder

1982 Mack Truck, 1000 GPM with 750 gallon water tank, "Engine 6"

REPORT OF THE FIRE CHIEF

To the Town Manager, Selectmen and Citizens of Farmington:

This year saw the number of calls drop thirty in number from the previous year, with chimney fires still receiving the largest percent, followed by smoke scares, structures, cars, false alarms, service calls, grass fires and mutual aid. The dollar value in property loss this year was the highest in several years. Two of the structural fires resulted in the loss of life. In April, two females, one a teenager, died in a house fire on High Street. In mid-September another house fire claimed the life of a young child. I cannot encourage the citizens enough to be careful with chimneys and wood stoves and to know where your children are.

Our new Mack 1000 G.P.M. Pumper was delivered in mid-April. This new piece of apparatus was tested and placed in service the first of May. With this new fire attack unit, we have a first class attack force.

A new fiber glass booster tank has replaced the cracked stainless steel tank on Engine 5 (Mack) and the pump on Engine 4 (Thibeault) has been rebuilt.

On behalf of the Officers and members of the Farmington Fire Department, I wish to express my sincere thanks and appreciation to Vertie Allen for the many long hours she has given to the Town and to the Fire Department.

Respectfully submitted,
Robert L. McCleery
Fire Chief

1982 Report of Fire Chief & Training Report

TRAINING REPORT, FIRE DEPARTMENT

To the Town Manager, Selectmen and Citizens of Farmington:

During 1982 the Department held 15 regular training sessions, either at Farmington or the Falls substation with an average attendance of 15 firefighters. This isn't as high as we would like to see but because of work schedules and other commitments we are limited in numbers.

We have made a continued effort to indicate to all firefighters the need to be well trained in fire fighting skills and methods. This is necessary for one's own protection as well as knowing the proper use(s) of today's equipment. The citizens who support the emergency units should, in turn, be given the best available and speedy assistance when necessary.

The subjects covered this year were as follows: Protective Breathing Apparatus uses and maintenance, C.P.R. refresher, Aerial Ladder and Ground Ladder uses, Hydrant Hookups and instruction, Treatment Plant rescue operations, Hose Testing and Hose use techniques, Salvage Covers and how to use them, Pumper Operations and Draft Pumping and an evacuation at UMF.

We had two men, Tim Hardy and Nelson Collins, attend the two week Fire Academy held in Presque Isle. This now gives the Department three members who have received this extensive training course. They are being used as instructors here in our Department. Several firefighters attended the Fire Attack School at Orono and the Franklin County Attack session. We encourage attendance at these sessions because many top-notch instructors present the materials.

During Fire Prevention week we held an open house here in Farmington and several citizens, with their children, came to look us over and observe a training session in progress. Several favorable comments were made about this presentation. We also placed a mannequin, dressed as a firefighter, in two food stores so that the young and old could see what a firefighter looked like. This was well accepted.

I would like the citizens of Farmington to know that they are always welcome to visit the Fire Department and call upon our services as needed. They should be proud of their Fire Department and the recognition it receives from other areas.

I take this opportunity to thank the men who have assisted in the training sessions this year. Their interest, support and dedication allow us to have an effective Training System.

> Respectfully submitted,
> S. Clyde Ross
> Assistant Chief

1982

REPORT OF THE FIRE CHIEF

To the inhabitants of the Town of Farmington, I hereby submit my annual report for the year 1983.

This was an average year for fire calls, with a very high fire loss in buildings and stock, totaling nearly a million dollars. There were 13 structure fires in the Farmington area with Starbird Home Care Center the largest. Several of these required mutual aid help from Temple on stand-by at Central Station mainly, and air bottles and man power from Wilton, with Chesterville or New Sharon to cover the Falls area. Farmington Fire Department responded to 12 mutual aid calls in Chesterville, Industry, New Sharon, New Vineyard, Strong, Temple and Wilton. Wilton and Temple fire departments responded to the aid of Farmington Fire Department 6 times this year. Some of the calls that the fire department responded to In 1983 were: chimney fires (45), vehicle fires (21), grass (10), false (13), smoke scares (9), accidents, assist Keegan Ambulance personnel, and assist extrication (7), and (1) wood fire in January. An assortment of miscellaneous calls (30) consisting of: electrical malfunctions, trees on power lines, stove and furnace blow back, gasoline and chemical leaks and vapors, vandalism, personal assistance, debris, cellar pumping and others were also answered.

In July, the third annual Franklin County Fire Attack School was held in Farmington with over one hundred in attendance.

In April, Farmington Fire Department was host to a pilot program sponsored by the Department of Education and Cultural Services, Fire Training and S.A.D. #9 Vocational Education Department. This program was to train fire department personnel on the Instructor I level to become certified fire instructors in their own and other departments. Twenty people attended this school of which six were from Farmington Fire Department.

In August, four Farmington Firemen attended the two week Maine Fire Academy in Presque Isle.

In October, Farmington hosted another school by the same sponsors, the Fire-Fighter I school held for the first time (outside the Fire Academy) with an enrollment of twenty-four, five from this department.

These schools have been very helpful and profitable to this Fire Department in adapting our people and those previously trained at the academy in carrying forward the vigorous training programs outlined by the training officers of the fire department. At the year end, Farmington Fire Department has seven firemen trained in the competences of Fire-Fighter II and five firemen trained in the competences of Fire-Fighter I.

An H.K. Porter Extrication Tool, complete with spreader and cutter, along with three air bags of different sizes were added to the rescue equipment this year. This tool is operated from air bottles and has a ten ton rating. The tool was used twice in November to free accident victims trapped in cars.

Tours and pre-fire plans have been and will continue to be made of the hospital, nursing care facilities, housing complexs and new develop-ments.

As more residential and commercial developments are constructed both in and outside the hydrant area with high property value, it places a greater demand on the fire department for added protection requiring more large diameter hose, portable tanks and tank trucks.

I would recommend to the Town of Farmington that within the next two years a request be made for a reevaluation of our fire protection classification to up-grade our present rating of 5C.

At the annual meeting in October, I was elected to a one year term as president of the Maine Fire Chiefs Association.

Respectfully submitted,
Robert McCleery
Fire Chief

1983 Report of Fire Chief & Training Report

FIRE DEPARTMENT, 1975 TO 1983

After the legislative sessions in 1974, the administration of the Town realized it would be necessary to establish a municipal fire department or contract with a private agency for fire protection. Prior to this time it was basically understood that Farmington Fire Department #1 was the fire control unit.

The Town Meeting of March 1975 voted to accept the Farmington Fire Department #1 as a municipal unit and enacted the Fire Department Ordinance. This contains rules and regulations along with stated operating procedures, which have been amended as conditions have warranted. Also at this meeting the Town voted to purchase a new Mack Fire Truck with a 1000 gallon per minute pump. The total price of the unit was $53,250. (This was some different than the purchase of Farmington's first unit in 1860 for $400.) It was also voted to expend $807,380. for a new municipal/fire station building on Lower Main Street. The new Mack Truck arrived in December of that year. It was really a great boost to the fire fighting capabilities of the Town.

December 13, 1976 saw the Fire Department move from High Street to Lower Main Street. As some of you remember, the Fire Station had been located on High Street for many years, either for horse drawn units or the new gasoline units.

In 1977, Chief Forest L. Allen resigned that position after serving the Town for 34 years as a fireman and officer. Robert L. McCleery was appointed the new Chief in April of 1977 and Glenwood Farmer was named Deputy Chief. Richard Russell was appointed Assistant Chief giving the department a full slate of officers.

As the demands upon the Fire Department increased, not just for fire emergencies, it became necessary to divide the department into Engine Companies and appoint additional fire officers. Since 1978, the following have been selected or promoted: Clyde Ross, Assistant Chief - 1978; John O. Bell and Harold Hemingway, Captains - 1978; Clyde Meader, Lieutenant - 1978; Kenneth Durrell and Morrell Collins, Lieutenants - 1980; Mahlon Moore, Lieutenant - 1983; and Terry Bell, Lieutenant - 1983, Assistant Chief - 1984 (Richard Russell resigned). These officers have helped in updating our training program which has become more extensive in recent years.

Training for the Fire Department over the past few years has seen marked improvements in the use of self contained breathing apparatus, pump and hose lays or uses, interior fire attack/ventilation, C.P.R., ground/aerial ladder practices and public service and prevention. As the citizens call for more assistance, it is necessary to upgrade fire personnel and equipment.

Along with our own training program, the State of Maine Fire Service office has offered several courses with very capable instructors. We have in the department seven Fire-fighter II, five Fire-fighter I and six certified Instructor I people that are being used to help other firemen here in our department. These above men have either taken the State course in Presque Isle (130 hours) or here at Farmington (F.F. I - 64 hours and Instructor I - 36 hours). We have incorporated the competencies and skills as set forth by the International Fire Service Association in our training. This is pretty much accepted nationwide as the standard for fire fighting.

Our long range training program will hopefully see more of the fire fighters certified as F.F. I or F.F. II, cold water rescue training, expand training with Porter Extrication Tools, building ventilation, hazardous materials, pre-plans and fire prevention programs. We are making plans for a more extensive training program with the mutual aid departments in neighboring towns. They are a valuable asset to us because of their trained personnel and equipment.

1983

Robert L. McCleery 275

After many years of waiting and planning, an Aerial Ladder Truck was purchased in 1978. This unit was delivered early in 1980. We certainly are pleased with the support given for this unit by the Town and the University of Maine at Farmington. The aerial ladder unit has been a great help in meeting emergencies in multistory buildings.

During 1980, a replacement squad unit was purchased to improve our line of vehicles and update our capacity to meet public needs. The old one had served us well but had some rather extensive problems.

In today's fire attack, S.C.B.A. (self-contained breathing apparatus) is being used in practically every structure fire. Air for these units has become an item of expense and concern. To assure ourselves of an adequate supply of needed air, we purchased a compressor system for filling our own tanks. This past summer we added a "Cascade System" for greater storage and easier tank filling.

In 1974 - 1975, the Insurance Service Orgainsation made a survey of the Fire Department and water supply. The Town soon received an updated and improved I.S.O. rating. Among the recommendations made were purchase of an aerial unit, improved training program and regular rotation purchase of new fire apparatus. In keeping with these suggestions, we are currently looking at a rotation of trucks every ten years. This simply means that with three "front line" pieces, they would have about a thirty year service span before a replacement would be ordered. Our most recent unit was the new Mack Truck purchased in 1981. Engine #3 was then placed at the Falls Fire Station. Long range plans have been worked on by the department and Town Manager.

Like every service organization or emergency unit, the costs of operation have shown an increase. Our operating budget in 1975 was $41,970.00 and $89,330.00 in 1983. This may appear to be substantial but the number of calls and updated equipment purchased is a large part of the increase.

Since the purchase of the Porter Extrication Tool this past summer, we have designated a rescue unit from department personnel. We are in the process of training our people in the uses of the tool and special equipment. The Porter Company has sent a training representative to assist local instructors with this unit.

In 1975, the Town also accepted the Falls Fire Company as a part of the Town Fire Department and has received prompt and capable service from its members. They are truly a part of our department, participating in training, internal affairs, equipment and building upgrading, and fire fighting.

In 1982 Vertie Allen, who began dispatching for the Fire Department in 1969, retired. She performed one of the most important public services in our community with a dedication and consistency that is rare indeed. Since December 1982, the Franklin County Sheriff's Office has dispatched our fire services, in a very commendable manner.

In closing I would like to thank the members of the Fire Department for their continued involvement in serving the needs of the citizens of our community. The Town should be proud of its efforts in supporting the Fire Department, and the tremendous gains made in the last several decades.

Respectfully submitted,
S. Clyde Ross
Assistant Chief

1983

REPORT OF THE FIRE CHIEF

To the inhabitants of the Town of Farmington, I hereby submit my annual report for the year 1984.

There were more calls this year than any year since I have been Chief, with less serious property damage and time involved. There were 11 structure type fires in Farmington resulting in approximately $150,000. loss. Some of these required mutual aid help from Wilton, Chesterville and Temple. The Fells Sub Station responds to Farmington on all structure fires. Some of the calls Farmington Fire Department responded to were: Use of extrication tool (8 times since it was put into service in November of 1983), chimney fires - the largest percentage of calls (54), vehicle fires (11), grass and brush (6), false alarms and system malfunctions (15), smoke scares and related problems (19). An assortment of miscellaneous calls consisting of electrical malfunction, trees on power line, gasoline and chemical leaks or spills, debris, service calls, cellar pumping and others.

This year the Fire Department has filled 209 air bottles for Farmington and 333 for other towns. Several chimney and stove installation inspections have been made. Also, inspections of new and existing businesses as well as apartment house renovations and new construction.

Throughout the year, programs on fire prevention have been carried out in the community, and our schools. In July, the fourth annual Fire Attack School was hosted by Farmington Fire Department with 100+ in attendance.

In October several firemen completed their Fire Fighter II training, with acompanying Basic First Aid, and were certified Fire Fighter II s.

An extensive in-service training program, under the direction of Fire Prevention and Training Officer Assistant Chief S. Clyde Ross, has been carried out. This consists of classroom and hands on training in pre-planning, fire behavior, use and care of protective breathing apparatus, maintenance of pumps and pumping operations, ladder use and care, hose handling, adaptors and appliances, hose testing, use of extrication tools, portable tank and tanker shuttle operation and forest fire simulation. Pre-planning and in-service training will continue in the coming year as well as inspections and tours of existing and new complexs.

1984 Report of Fire Chief & Training Report

As the Town grows and expands in both residential and commercial areas, a greater demand is placed on the Fire Department. A report of Fire Department proposals, projections and replacements over a period of years has been presented to the Town Manager and Board of Selectmen. This report projects the replacement of vehicles and equipment over 10 years.

As the community continues to grow some consideration will have to be given to full time firemen. Employers are becomming more reluctant to release their help to respond to fire calls, some will not do so even now. There is a need for inspections by the local Fire Department of new construction, remodeling of apartment and rooming houses, day care centers and homes, buinesses and places of amusement. These inspections are not being carried out by this department due to the lack of personnel.

Respectfully submitted,
Robert L. McCleary
Fire Chief

1984

During the past year, the Fire Department has been busy with its training program as well as assisting the public with its emergencies. These training sessions have included such topics as tanker water shuttles, protective breathing apparatus, ground and aerial ladder operations, forestry simulation fire attack, salvage and overhaul, hose handling and adaptors, hazardous materials and the K.K. Porter extrication tool. Most of the sessions were conducted by our own trained department instructors. This is where we now begin to benefit from the Instructor I and Fire Fighter I and II Academies that our people have attended in the past years. A total of 20 sessions were held, which means 600 man-hours of training in the department.

Again this year, Farmington hosted the 4th Franklin County Fire Attack School and 115 people spent the day in fire fighting skill development classes. Later in the fall, the Fire Fighter II Academy was held here and 19 county people completed this course. To become certified F.F. II personnel, all who had not taken a first aid course prior to the time, then attended and completed this last requirement. Our department should be very pleased and proud of the efforts that have been shown by its firefighters.

This year during Fire Prevention Week, department personnel were in the public schools of the Town and were able to talk with more than 550 elementary students. The major topic was home safety, but also what a fireperson does and how he/she is trained was of great interest to the young people. They really have given serious thoughts to this subject, safety. U.M.F. again helped us with their participation in an evacuation drill with live action taking place. It caused some talk and disruption about the campus, but it was a successful venture for all.

The department has again been able to work with the hospital and several nursing care facilities with mock drills and emergencies. We have all learned some valuable lessons during the sessions and the critiques that have followed.

It might be of interest to note that the department's training committee has met several times to set up an agenda or schedule for training topics and get the necessary instructors (many are now from our department). We have adopted the National Fire Prevention Associations training program as our guide. If we are going to train, then let it be to a goal that is recognized nationally..

In closing, I wish to thank the instructors and firefighters who have been so faithful in their responsibilities to the citizens and the department. The Town should be proud of their firefighting personnel and of their willingness to support the department's programs.

Respectfully submitted,
S. Clyde Ross

1984

REPORT OF THE FIRE CHIEF

To the inhabitants of the Town of Farmington, I hereby submit my annual report for the year 1985.

This has been a busy year, more calls for various incidences. Eleven structure fires resulted in an approximate dollar loss of $255,500. Other calls responded to by the Fire Department were chimney fires (45), vehicle fires (17), accidents and/or extrication calls (9), grass, brush (9), debris, trash, smoke, furnace, stove, C.M.P. line problems, electrical and gasoline spills (32) false alarms (29), miscellaneous and service calls (13). The Mutual Aid cooperation between the several surrounding towns continues to work satisfactorily. We have assisted and have also received assistance as needed.

From January through December, the Fire Department filled 926 air bottles for Farmington and neighboring towns.

Apartment house inspections are continuing and are top priority, as well as the other building inspections we are called upon to perform. These inspections will continue and will probably increase as time goes on.

Fire Prevention Programs have been carried out in the community and especially at the local elementary schools under the direction of Assistant Chief S. Clyde Ross and several other fire fighters. I wish to express my thanks to the personnel of S.A.D. #9 for granting "time off" from teaching duties to accomplish these programs.

Extensive in-service training is continuing under the direction of the training officer and department instructors as you will see in the Training Report. Again in July, the annual Franklin County Fire Attack School was hosted by the Farmington Fire Department. The yearly turn out for this event is close to 125 fire fighters.

On behalf of the Fire Department, I wish to thank the Ladies Auxiliary for their back-up with food and beverages at the fire scenes, to the Keegan Ambulance Service for their support and to Chief Smith, Sergeant Wilcox and the Farmington Police Department for the support at the fire scenes with traffic control and the process of investigations. I also thank the Franklin County Agricultural Society for the use of the grounds and buildings used by the Fire Department.

On September 27, 28, and 29, Farmington Fire Department hosted the 32nd Annual Maine State Federation of Fire Fighters Convention. This was a first for our department. There were several thousand people along the parade route to view the units in the procession that had assembled from across the State. Considering the threat of Hurricane Gloria, the convention was a great success. I wish to thank all the citizens who supported this project with their time or funds.

In conclusion, the Fire Department thanks the citizens of Farmington for supporting our efforts and the needs and programs it represents.

Respectfully submitted,
Robert L. McCleery
Fire Chief

1985 Report of Fire Chief & Training Report

FIRE DEPARTMENT TRAINING REPORT FOR 1985

During the year of 1985, the Fire Department has held 16 training sessions, either at the Farmington or Farmington Falls station. The average attendance has been 13 per session. The length of the sessions varied from one to three hours. We try to keep within the two hours time if possible.

The training sessions have covered such topics as Self Contained Breathing Apparatus, Hazardous Materials, Pumper Training, Ropes and Knots, Hose Testing, Rockwood Foam Demonstration, CPR Refresher, Disaster Drill and U.M.F. Scott Evacuation Drill. These don't sound like very impressive titles or activities but the instructors and real-life situations have made them very worthwhile.

The session on Ropes and Knots was conducted by persons from the Rumford based Army Mountain Rescue unit, a new branch recently introduced in this area. The materials and hands-on experiences were extremely helpful to our people. Later in the spring, we had a second session on Ropes and Knots.

Franklin Memorial Hospital Director of Plant Operations, Peter Ross, invited us to attend a session on Hazardous Materials that one might find at the local hospital. We were given a brief description of the materials and shows the location for its storage. This was a joint session with the Police Department. It was very helpful to all of our people.

On three different occasions we worked with the Keegan Ambulance Service in conducting training that would help both groups. These were in evacuation and use of emergency equipment used at accident scenes. The CPR Refresher course was also conducted by the above service. Ten of our fire fighters completed this course and were re-certified.

Again this year the department participated in Mutual Aid Training with our neighboring towns. This training was done in Tanker Water Shuttles, Class "A" (structural fire attack) Burns, Self Contained Breathing Apparatus and the Rockwood Foam Demonstration. This is a vital part of our program today as departments are calling for assistance and need qualified personnel at fire and other emergency situations. Mutual Aid has served our Town and adjoining towns very well in the past years.

The use of 4" large diameter hoses are increasing in our area. We have held special schooling on the laying and pumping of water over a long distance, up to a half mile. This is no easy task as communications and pump pressures are vital to the success of this operation. There is still work to be done in this operation.

The work of the Training Committee is never done and the members of the committee have met and planned the program for 1986. It is only through the continued support of the Town and its' citizens that this program can continue. It has paid off for all of us in one way or another. At this time, I would like to thank all citizens who have assisted in our training during 1985.

Respectfully submitted,
S. Clyde Ross
Assistant Fire Chief

1985

REPORT OF THE FIRE CHIEF

To the inhabitants of the Town of Farmington, I hereby submit my
annual report for the year 1986.

Another year has passed with over 200 calls for assistance.
Starting in January with the Tom Williams farm on Route 27 then
the Larry Tracy home on the Season Road in February. March saw
three severe structure fires; namely, the Luce house on Lower
Main Street, Sweetser's Mill on Route 43, and the Kolreg mobile
home on the back Temple Road. In May the Robert Hunter home by
the green bridge in Farmington Falls was destroyed. These were
the most serious of the structure fires, resulting in over
$200,000, damage.

Other calls to the Fire Department were chimney fires (41),
vehicle fires (8), accidents and/or extrication calls (12), grass
& wood (8), debris & dump, tires (16), false alarms or alarm
malfunctions (15), smoke, stove and furnace problems (14),
electrical related (6), spills and vapors (10), cellar pumping
(13), and service calls (15).

During the flood in January, the Fire Department with boats,
rescued 6 people stranded at the Dinner and 7-11.

The Fire Department has received mutual aid several times
this year, and also have responded to our neighboring Towns
several times. The most serious and destructive out of Town fire
this year was the Davis Farm in September.

Several inspections of apartment houses, businesses,
chimneys and other buildings have been made and will continue.

In the not to distant future, the Town should consider some
full-time firemen to adequately perform the needed inspections,
vehicle maintenance, pre-planning and record keeping.

Fire Prevention Programs have been carried out in the
community, at the fire station and the elementary schools under
the direction of Assistant Chief S. Clyde Ross and several fire
fighters demonstrated equipment. Again, I wish to express my
thanks to the personnel of S.A.D. #9 for granting time off from
teaching duties to accomplish these programs.

In reading the report of the training officer, you will see
that an extensive in-service training program is being carried
out by our several qualified instructors. The firemen, with
money they earned from the Fair Booth and Chicken Barbecue,
purchased a new Poseidon Air Compressor to fill the Cascade Air
System used in filling air bottles for self contained breathing
apparatus. This new compressor replaces an older, outdated model
and fills air bottles to the desired pressure quicker and more
accurately. Six hundred and seventy-two air bottles were filled
from January through December.

On behalf of the Fire Department, I wish to thank Police
Chief Smith and the Farmington Police Department for their
support in handling traffic and with investigations, also the
citizens. Farmington for supporting our efforts, needs and
programs.

Respectfully submitted,
Robert L. McCleery
Fire Chief

1986 Report of Fire Chief & Training Report

FIRE DEPARTMENT TRAINING REPORT FOR 1986

1986 saw no particular changes in the training sessions held at the Fire Department. As in past years, the importance of training, familiarity with standard equipment and new purchases, retaining man's skill and competency levels and adjusting to the new attack techniques have been stressed to the older fighters and our new recruits. The length of the training sessions have been varied, some one hour and others up to three hours.

Topics covered during the 1986 sessions were Self Contained Breathing Apparatus, Draft and Relay Pumping, C.P.R. Refresher, practice with the extrication tool and Car Fire Suppression, Ground Ladders and Aerial Ladder uses, Job Safety, Class "B" Burn, Rural Pitch and Fire Ground Survival. The latter three sessions were held in conjunction with over continuing Mutual Aid Training. New Sharon was the host for these sessions. Instructors for several of our schooling sessions have come from our own qualified personnel. Previous reports will indicate when and where these people received their credentials.

Forceable Entry and Fire Suppression tactics received much attention this past year as new methods and improved safety practices have been stressed. Injuries sustained in other departments across the nation in the above procedures have prompted reviews and new approaches. Hopefully we can keep abreast of these methods of improved attack.

This year there were no department members attending the Fire Academy. Also there was no Franklin County Fire Attack School held as had been the practice in previous years.

As in past years, the Fire Department has held training sessions with Keegan Ambulance Service and the University of Maine at Farmington. These sessions are well attended and are very beneficial to the organization involved.

The Fire Department Benevolent Association purchased a new air compressor and the members have been instructed in the proper procedures for filling S.C.B.A. tanks. This new equipment has made it possible to fill our air supply containers more efficiently.

The Camp with No-Burn Fire Prevention Program was again presented in the local elementary schools, grades 1 thru 5. This year 478 students participated in the instruction as did their teachers. The department's Training Committee has worked very hard to continue this valuable program with our young people.

Lastly, I again wish to thank the fire instructors, fire fighters and the citizens of Farmington for their continued support of the training programs.

Respectfully submitted,
S. Clyde Mose
Assistant Fire Chief

1986

Robert L. McCleery 283

REPORT OF THE FIRE CHIEF

To the inhabitants of the Town of Farmington, I hereby submit my
annual report for the year 1987.

Another busy year has passed with a record number of calls.
As each year passes, there is more time required of Fire
Department personnel in fire related incidents, calls for
assistance in other emergencies and community activities.

Some thought should be given to some full time firemen in
the not to distant future. These firemen could do inspections,
vehicle maintenance, preplanning and public relations.

Structure fires (15) resulted in approximately $195,000.
damage this year. Chimney fires resulted in the largest number
of calls, but have been steadily dropping over the last 3 years,
from a high of 65 in 1980 to 34 in 1987. Other emergency calls
responded to by the Fire Department were vehicle fires (16),
accidents and extrication calls (16), grass and brush fires (5),
dump and debris fires (11), false alarms and system malfunctions
(17), smoke and furnace problems (15), electrical related (6),
gasoline spills (9), mutual aid (13) and a service or
miscellaneous (71).

The Fire Department was very much involved the April 1st
flood of 1987. This included rescue of people by boat from
businesses and homes on Intervale Road. An attempted rescue of a
person at Farmington Falls resulted in a near tragedy for two of
our men at Farmington Falls. Numerous cellars and the Treatment
Plant had to be pumped out by the Fire Department, 36 in all.
The mutual aid corporation between the towns continues to be
beneficial to all, both ways.

Fire prevention programs are being carried out in the
community and in the schools under the direction of Assistant
Chief S. Clyde Ross and the committee. For this I wish to
express my thanks to the administration of S.A.D. #9 for granting
time off from teaching duties to accomplish these programs.

The Fire Department is asking the voters of the Town to
appropriate money for a new Aerial Ladder Truck. Due to the age
and unsafe condition of the ladder on the present Aerial Ladder
Truck, I feel that the only feasible way to go is for a complete
new Aerial Ladder Truck.

On behalf of Farmington Fire Department, I wish to thank
Sheriff Pete Durrell, Lieutenant Lee Dalrymple and other
emergency calls for the Towns in Franklin County. I wish to
thank Police Chief Sheridan Smith, Sargeant Investigaor Nolan
Wilcox and the Farmington Police Department for their support in
handling traffic and investigations; also the Ladies Auxiliary
for food and beverages when needed, and the citizens of
Farmington for supporting our efforts, our needs and programs.

Respectfully submitted,
Robert L. McCleery
Fire Chief

1987 Report of Fire Chief & Training Report

Fire Department Training of personnel is one of the major concerns that all leaders in the fire service are concerned about and spend much time administrating. The need of qualified fire fighters is an important part of any fire department. Our efforts continue to be directed in making each person comfortable in doing the tasks that are assigned to them.

This year the department has held 24 training sessions. The topics covered have included forestry techniques, SCBA, extrication tactics, water rescue methods, "class B" burns, aircraft crashes, ground ladders, relay pumping and driver training. These sessions were conducted by members of the local department and there were a couple guest instructors brought in for special sessions. The attendance at these sessions has been very good and time spent at the sessions is generally two hours.

The Franklin County Fire Attack School was again hosted by Farmington and 115 fire fighters from the county attended this one day session. It was mentioned that this was one of the best schools held in years. This speaks well for our instructors and those who participated.

There is one area that continues to be a problem in the training program and that is hazardous materials. We are aware that today there are many vehicles traveling over the highways and the materials being carried are to numerous to mention. How to handle a spill or leak is vital to the environment and all citizens in the area. This topic and training is being addressed at the state, county and local levels. We are involved and will continue to educate our people as quickly as possible.

We have held training sessions with the U.M.F. and Keegan Ambulance staff this year as in the past. It is necessary to work together and develop the good understanding in all organizations served by emergency units.

The Fire Prevention Committee has continued its work with the local schools and this year 485 students were involved in the programs. This is a fine opportunity for youth and adults to work and share a common idea, home safety. The students have been very cooperative in all sessions and visits to the fire station.

I want to thank the Training and Prevention Committees for their continued support and dedicated service to the citizens of Farmington. The citizens should also be commended for their support of all our programs.

Respectfully submitted,

S. Clyde Ross

Assistant Fire Chief

1987

Robert L. McCleery 285

To the inhabitants of the Town of Farmington, I hereby submit my annual report for the year 1988.

This has been a relatively quiet year as far as fire and emergency calls go, down eighty (80) called from a high year in 1987. This is due partly to the fact that there were less chimney fires, 34 in 1988, to a high of 55 in 1980. Also more citizens have gone back to oil or some other type of heat. Those that are burning wood now have better heating appliances, new chimneys and are doing a better job with dry wood.

The ten (10) serious structure fires in the Farmington area resulted in a property loss exceeding $181,000.00. The most serious problem this year was false alarms or system malfunctions, 34 in number. Most other emergency calls were well below the previous year. The other calls, service or miscellaneous, ran about the same.

This year I am again asking the Town of Farmington for some full time firemen. These firemen would do badly needed maintenance on apparatus and equipment, help with building inspections and follow up visits, preplanning, record keeping, service calls and some fire calls.

On November 15, the Farmington Fire Department received delivery of our new "Emergency One" 110' Aerial Ladder Truck, better known as "LADDER 1". Upon receipt, training was started immediately and will continue throughout the year.

On December 4th, at approximately 1:90 A.M., an accident occurred at the Fire Station, involving a privately owned vehicle and two (2) Fire Department vehicles. A car being pursued by a police car drove through one of the overhead doors to the Fire Station, striking the Squad Truck pushing it back into the old aerial truck, then coming to rest on the front tire of the new ladder truck. This accident caused more than $19,000.00 to these vehicles and to the station.

As in the past, Assistant Chief S. Clyde Ross has carried on extensive Fire Prevention Programs in the community and in the schools. For this I wish to express my thanks to the administration of SAD 49 for granting time off from teaching duties to accomplish these programs. Another aspect of our training program is the construction of a "Burn Building" at the Landfill for the training of firemen in interior fire attack. The Fire Department has a "Smoke Trailer" in operation for the purpose of training firemen in the proper use of Self Contained Breathing Apparatus (SCBA).

On behalf of the Farmington Fire Department I wish to thank the Fire Dispatchers at the Franklin County Sheriff's Office for the efficient dispatching of our calls. I wish to thank Chief Smith and the Farmington Fire Department for their continued help with traffic and investigations; and also the people of the ambulance service, Ladies Auxiliary, and all others who have so faithfully assisted us.

Respectfully submitted,

Robert L. McCleary, Fire Chief

1988 Report of Fire Chief & Training Report

REPORT OF THE FARMINGTON FIRE DEPARTMENT TRAINING FOR 1988

The Farmington Fire Department has again continued the
extensive practice of inservice fire training and up dating of
suppression techniques. During the past year there have been 21
Training Sessions, most of them lasting at least two hours. We did
have three day sessions and these were for Mutual Aid assistance,
Cold Water Rescue and Auto Extrication (the latter two were
taught by the Delta Ambulance personnel). The hands on items have
really worked well for us this year. The more we can instill
confidence in one's own abilities, the more effective the person
is in the variety of responsibilities given.

The topics we have used in the training this year were
S.C.B.A., Arson, ground ladders and the use of the Aerial Ladder
unit, protective turn-out gear, ventilation techniques, fire
ground survival and a refresher in CPR. These sessions were well
attended and beneficial to all that attended.

The Franklin County Fire Attack School was again held in
Farmington and only 53 persons attended this year. This was a
much lower turnout than expected and the future of this school is
in doubt. Many departments are now doing their own training and
the county school may have served its last session.

Again this year the Fire Prevention Week activities included
visits to the Fire Station by the first graders at the Mallett
School and several nursery and pre-school groups. These young
people were accompanied by teachers and parents. Contact was also
made in individual classrooms at the Mallett and Ingalls Schools
by members of the committee serving the prevention needs. Four
hundred twenty-five students were involved in these meetings.
The main topic was "what I can do to make my home and family
safe the whole year."

Several of the officers have attended State sponsored
seminars in Waterville, Augusta, Freeport and Bangor during the
past year. These have helped the department in a variety of ways.
New ideas have been obtained and new contacts made that will be
of importance in the months ahead. One of the programs attended
was the Juvenile Fire Setters seminar in Augusta. This program
deals with youth who are playing with fire and causing extensive
damage to properties in their communities.

Hazardous Materials are again receiving much attention to
the fire service and each first responder is required to have
twenty-four hours of training in related subject areas. We have
fulfilled that requirement in our programs.

In closing, I just want to again thank all the training
committee members and the prevention committee for doing an
outstanding job this year. We have served many in our community,
young and old alike. The Fire Department thanks the citizens for
their support and understanding the needs of the fire service and
its programs.

Respectfully submitted,

S. Clyde Ross
Assistant Fire Chief

1988

REPORT OF THE FARMINGTON FIRE DEPARTMENT

To the Inhabitants of the Town of Farmington, I hereby submit my
annual report for the year 1989.

This has been another relatively quiet year as far as fires
go. Chimney fires are down, citizens are burning less wood and
using other types of fuel for heat.

A summary of some of the calls the Fire Department has
answered this year are: structure fires - 10 with a property loss
exceeding $157,500.00 (3 totaling more than $30,000); chimney - 32;
vehicles - 10; accidents and extrication - 11; grass, brush and
dump - 20; false alarms and system malfunctions - 26; smoke,
stove and electrical problems - 19; mutual aid, spills, service
and miscellaneous - 47. Farmington Fire Department also received
mutual aid from Chesterville, New Sharon, Industry, Temple and
Wilton. As Fire Chief I extend my sincere thanks for their excellent cooperation. There is an excellent mutual aid response
between the area towns. This has come about through the upgrading
of equipment, training and a better understanding of each other's
needs.

This year a new radio console and dispatching equipment were
purchased by the county to upgrade their dispatching
capabilities. The area fire departments purchased a new base
station and tied into the new console at the county dispatch
center (jail). At present there are nine (9) Fire Departments
being dispatched by the county.

In August, Farmington Fire Department purchased a new Ford
F350 Squad Truck to replace the one that was lost due to the
accident of December 4, 1988. This purchase was made possible
upon receipt of the insurance money from the accident, the
appropriation by the town, and by a donation from the Firemen's
Benevolent Association.

In-service training in all aspects of fire suppression,
hazardous materials identification and fire prevention have been
carried on by certified instructors from our own department.
Again this year, Assistant Chief Clyde S. Ross has carried on
extensive Fire Prevention Programs in the community and in the
schools. S.A.D. #9 generously grants time for these programs, as
it is to their benefit as much as it is to the fire department.
Teaching safety begins with the young. For this I wish to express
to them my thanks.

On behalf of the Farmington Fire Department, I wish to thank
Chief Nolan Wilson, Former Chief Sheridan Smith and the
Farmington Police Department for their continued help with traffic control at fire scenes. I wish to thank the fire
dispatchers at the Franklin County Dispatch Center and also the
people of the ambulance service, Ladies Auxiliary and all others
who have so faithfully assisted us.

                                    Respectfully Submitted
                                    Robert L. McCleary

1989 Report of Fire Chief & Training Report

REPORT OF THE FARMINGTON FIRE DEPARTMENT TRAINING FOR 1989

To the Town Manager, Selectmen and Citizens of Farmington:

During the past year the fire department has been actively involved with the in-service training of fire department members. As in the past years, we have put greater emphasis on building individual skills and confidence in one's own abilities as the team concept is developed in fire suppression. This seems to be the method used today in facing the emergencies directed to our attention.

Topics covered this year were similar to those in the past namely the use and care of SCBA (this was a full course of 13 hours taught by our own qualified instructors), the Aerial Ladder truck (uses, location of the ladder at fire scene, operation of the vehicles itself), the annual testing of the fire hoses, Haz Mat update, ropes (uses and knots useful to the fire service people), CMP safety program as presented by their safety committee, CPR refresher course, Radon in the home and how it is causing concern, Class "B" burn (flammable liquids), a tour of a local printing business and building construction practices. In addition to our local training session, several persons have attended the regional Fire Fighter 1 Academy, area fire attack schools, state sponsored seminars, National Fire Academy State of Maine weekend study package and special programs that have been presented at the County Firemen's Association meetings. There is a wide variety of programs being offered today, and we are trying to take advantage of as many as possible. This requires many hours of an individual's time, and there are limits as to how many hours one can spend in this business.

During Fire Prevention Week in October, department members presented programs at the local elementary schools, two nursery schools, Franklin Memorial Hospital and provided tours of the fire station to interested youth groups. It was a very busy week for our members. It is a good feeling to know that the citizens are interested in fire safety and are willing to request programs. We will do all we can to make our people available to assist with this important function.

In closing, I wish to thank all the members of the training committee and the prevention committee for their continued interest and support of the necessary programs for the department and the community. To you the citizens, thanks for your understanding and patience as we have moved apparatus, generated smoke and asked for support in funding the programs.

Respectfully Submitted
S. Clyde Ross
Assistant Fire Chief

1989

REPORT OF THE FARMINGTON FIRE DEPARTMENT

To the Inhabitants of the Town of Farmington, I hereby submit my
annual report for the year 1990:

Starting in December of 1989 through November of 1990, the Fire
Department has been very busy with 194 emergency calls, and fifteen
structure fires with a property value loss of $305,000.00.

Every year, the Fire Department responds to chimney, grass, brush
fires, accidents, extrication calls, false alarms, system
malfunctions, smoke, stove and electrical problems, oil and gas spills
and a variety of miscellaneous and service calls.

Farmington Fire Department and the surrounding towns have an
excellent Mutual Aid response between them and the cooperation is
excellent.

In August of this year, Farmington Fire Department was involved in
it's first Hazardous Material spill involving a tank truck loaded with
6000 plus gallons of Black Liquor. Our mutual aid people responded to
assist, along with International Paper Co., Department of
Environmental Protection and RST Hazardous Material Team. All things
considered, the evacuation, salvage and overhaul went very well.

This year, the old aerial truck was converted to a tank truck,
with all the work being done locally. This truck is a self contained
unit carrying 1800 gallons of water, 1 portable tank, a 300 G.P.M.
pump, ladder, generator and a portable pump. It will be a very useful
piece of equipment for the Town of Farmington.

In Service Training is an on going part of our operation and is
carried on under the direction of Assistant Chief Clyde Ross, using
Certified Instructors from our own department. Also fire prevention
programs are being carried out community wide in the SAD #9 schools.

In December of 1990, the 9-1-1 emergency number went into effect
replacing the old 778-2120 fire number. This covers all 778 prefix
numbers and is for all ambulance, fire and police emergency's.

Once again, on behalf of the Farmington Fire Department, I wish to
thank Chief Nolan Wilcox and the Farmington Police Department for
their continued help and support with the traffic control and
assistance at fire scenes. Also, the dispatcher's at the Franklin
County Dispatch Center, Delta Ambulance crew, Ladies Auxiliary and all
others who have assisted the Fire Department.

Respectfully Submitted,
Robert L. McCleery
Fire Chief

1990 Report of Fire Chief

REPORT FROM THE FIRE DEPARTMENT

To the Selectmen, Town Manager
and Citizens of Farmington:

In 1991, Farmington Fire Department responded to 189 emergency calls, 20 service calls, and participated in 23 training sessions for a total of 232. The in-service training program is under the direction of Deputy Chief Ross. He is assisted by other officers and members of the department. The detailed activities of the training program are in the training report, along with the activities of our Fire Prevention program. The Farmington Fire Department Cascade System filled 655 air bottles for our department and 870 air bottles for our mutual aid departments. The fire department responded to the usual chimney, structure and grass fires. We also responded to numerous auto extrication calls, alarm malfunctions, smoke, stove and electrical problems, gas and oil spills and a variety of miscellanious calls. The mutual aid between Farmington Fire Department and our neighboring towns is excellent due to outstanding cooperation.

All but two fire departments in the county are now on line with the 9-1-1 Emergency Dispatch System and are dispatched through the Franklin County Dispatch Center located at the jail.

Once again, on the behalf of the Farmington Fire Department, I wish to thank the Police and Highway departments and Delta Ambulance for their assistance at emergency scenes. I would also like to give a special thanks to the fire dispatchers, Ladies Auxillary of the various fire departments and all others who have assisted the fire department in the past year.

Respectfully Submitted,
Robert L. McCleery,
Fire Chief

1991 Report of Fire Chief

REPORT FROM THE FIRE DEPARTMENT

To: The Town Manager, Selectmen and the Citizens of Farmington,

Farmington Fire Department responded to 293 calls in 1992, 241 emergency situations and 52 service calls, the largest number of calls this department has ever responded to in one year.

Some of the calls were for: 13 structure fires with a total value of property loss of $181,000.00; 29 chimney fires; 9 vehicle fires; 52 accident and extrication calls; 18 grass, brush, and debris fires; 18 false or alarm malfunctions; 26 smoke, stove and electrical related fires; 14 gasoline and oil spills; 3 for assistance in search for lost children; 24 cellar pumping and flooding and broken pipes. The Fire Department responded to 28 Mutual Aid calls and received Mutual Aid 25 times. The cascade system filled 748 air bottles for Farmington Fire Department and 979 for Mutual Aid departments. Our Mutual Aid relationship with the neighboring towns is excellent and is a benefit to all citizens.

The Training and Fire Prevention programs, under the direction of Deputy Chief Ross are working very well as shown in the Annual Training Report.

Extensive Haz Mat Training was conducted this year with several firemen attending.

Once again, on behalf of the Farmington Fire Department, I wish to thank the Police, Highway Department and Delta Ambulance for their assistance at emergency scenes, also to the fire dispatchers at the County Dispatch Center. A special thanks to the Ladies Auxiliary and all others who have assisted the Fire Department in the past year.

Respectfully submitted,
Robert L. McCleery,
Fire Chief

1992 Report of Fire Chief

ANNUAL TRAINING REPORT FOR THE FIRE DEPARTMENT

1992

During 1992 the Fire Department had 25 training sessions on various subjects. They were as follows:

Haz Mat: The regular topic for the members and
the on-going extensive classes for the
technicians who are now at the
specialist level
SCBA Class and Hands On
Safety Laws for the Fire Service
Appliances and Tools
Hydrants and Proper Hook-ups
Forestry Application Skills
Ladders and Tools
Extrication
Stress Management
HIV Virus and Related Information
Hose Testing
Pumps and Relay Lines from Draft
Disaster Drill
2 Class "A" Burns: These were with Mutual Aid
Companies
Ladder Truck Operations and Orientation
Fire Department S.O.P. Regulations

These have been conducted by in-house and outside instructors. We are fortunate to have these people with the expertise necessary close by.

The on-going Fire Prevention Programs were used again this year. Department members went to the local schools and talked with 425 young people. The Mallett School 1st graders made their annual visit to the station with their teachers and some interested parents. There was an "Open House" here at the station and many citizens took time to visit and discuss areas of interest to them. We appreciate their interest and support of the programs.

Once again the local insurance companies supported the Fire Prevention Programs and offered materials to make the topics more thorough.

Department members have spoken to several service organizations presenting numerous topics and answering questions from the members present.

As most of you realize, there is a constant change in Fire Fighting Technology and Equipment that is available to fire fighters. We attempt to keep our people up to date with as much information as possible to help assist the citizens with any situations that might occur.

The continued support of the citizens is appreciated and if we can be of assistance to any of you, please come in or call us. Finally, it is your continued interest that keeps us going and able to address your needs.

Respectfully submitted,

S. Clyde Ross, Deputy Chief
Farmington Fire Department

1992 Deputy Chief's Training Report

REPORT FROM THE FIRE DEPARTMENT

To The Town Manager, Selectmen and the Citizens of Farmington:

The Fire Prevention and Training programs are working well under the direction of Deputy Chief Ross as shown in his annual training report.

As a part of the Town of Farmington's Bi-Centennial Celebration, the Farmington Fire Department will be hosting the 45th Annual Convention of the Maine State Federation of Fire Fighters, Inc., September 9, 10, and 11, 1994. The Fire Department participated in the Annual 4th of July parade and Chester Greenwood Day Celebration.

Farmington Fire Department responded to 271 calls in 1993; 234 emergency and 37 service calls. Some of the calls were for:

16	Structure fires
27	Chimney Fires
13	Vehicle Fires
49	Accident and extrication calls
4	Grass, brush and debris fires
33	False or alarm malfunctions
36	Smoke, stove and electrical related fires
9	Gasoline and oil spills
5	Miscellaneous

The Fire Department responded to 24 mutual aid calls and received mutual aid 19 times. The Cascade System filled 509 air bottles for Farmington Fire Department and 332 for mutual aid departments. Our mutual aid relationship with the neighboring towns is excellent and is a benefit to all.

The requirements for the work load in the Fire Department is increasing each year because of the unfunded requirements of the State and OSHA which require more training, more equipment and more paper work. Example: confined space rescue, air quality test for SCBA and general fire fighting activities.
Once again on behalf of the Farmington Fire Department, I wish to thank all those who have assisted us this year, especially the Police Department, Highway Department, the personnel of Delta Ambulance and the dispatchers at the County Dispatch Center. A special thanks to the Ladies Auxiliary for their time and effort in providing food and beverages in time of need.

Respectfully submitted,
Robert L. McCleery
Fire Chief

1993 Report of Fire Chief & Training Report

ANNUAL TRAINING REPORT FOR THE FIRE DEPARTMENT
1993

Now that the year 1993 has come to a close, it is time to reflect on the training that has been done in the Fire Department. This training is not just for the members, but it helps the department better serve the community and its extensive needs. Some of you in the public sector don't realize that today's fire fighting is more than just putting on gear and riding on the rear step of a fire apparatus (by the way, it is outlawed today by government mandate). It is necessary to update the skills and expertise of a fire fighter just as in any other profession. We have tried and continually seek new information to keep us abreast of the current standards and mandates that come across the Chief's desk.

Several of the topics covered and studied this year were very familiar to the Fire Service but necessary to meet regulations. They were SCBA, HAZ MAT, EXTRICATION, HOSES and TOOLS, RELAY PUMPING, VENTILATION, CONFINED SPACE RESCUE, WATER RESCUE, CPR, and tours of recently constructed buildings - WAL-MART, SHOP & SAVE, FMH AND OUTPATIENT BUILDING, and a drill at 80 Main Street (some of you saw and heard about this).

During Fire Prevention Week, we had the Open House and many citizens took time to visit the displays (some by local merchants), view the fire videos and ask questions about topics of personal interest. The local first graders and teachers visited again and had the opportunity to see a response call and alarm while on their visit. This really opened the eyes and many questions were asked afterwards. Firemen visited several classrooms and gave instructive materials of interest to students and teachers. We have a very find Learn Not to Burn Program going on in the elementary school. Thank you teachers for doing such a nice piece of work.

The Hepatitis B series of shots has been completed. Thanks to those citizens who donated their time to make this possible. We appreciate your continued involvement in our programs. All fire fighters were given Physical Examination opportunities, again a mandate.

I keep mentioning Mandates, but you citizens really need to keep up on the current regulations before second guessing why we do the things we do and why we ask for new equipment and money for funding. Drop around the Station someday in your spare time and get to know what is really going on here. You would be surprised at what you might learn.

Juvenile fire setting is becoming a serious problem for many departments in the State and I sincerely hope that all of you are doing everything in your control to prevent this from becoming a major issue in our community.

In closing, let me make it clear that Fire Training is a very necessary item in our program and one that requires a great deal of time. You as a Town should be very proud of your Fire Department and the expertise that it displays in some very critical conditions. Be careful before you judge these people, you may need their services sometime.

Once again, thanks to the citizens who have supported our programs and offered their talents and thanks to you who support us in our on-going needs. If you have questions or problems, give us a call and we'll be available to assist.

Respectfully submitted,
S. Clyde Ross
Deputy Chief and Training Coordinator
Farmington Fire Department

1993

REPORT FROM THE FIRE DEPARTMENT

To: The Town Manager, Selectmen, and Citizens of Farmington:

I hereby submit my report for the Fire Department for the year
1994. As this was the Town of Farmington's Bicentennial Year,
the Fire Department was involved in three major events. The 4th
of July parade, one of the largest in many years; on September 9,
10, and 11th, the Fire Department hosted the 31st Annual
Convention of the Maine State Federation of Fire Fighters at the
Franklin County Fairgrounds. This three day event drew one of
the largest gathering of firemen and fire equipment ever
assembled, and the parade one of the longest with over 280 units
lasting over 2 hours. Saturday, December 4th, the Fire
Department participated in one of the largest and best Chester
Greenwood Day parades in many years.

The Fire Department responded to 290 calls in 1994; 231
emergency and 34 service, also 25 training sessions. Some of the
emergency calls responded to were: 18 structure fires, 23 chimney
fires, 20 vehicle fires, 44 accident and extrication calls, 12
grass, brush and debris fires, 25 false or alarm malfunctions, 26
smoke, stove and electrical related fires, 15 gasoline and oil
spills, 24 miscellaneous. Also mutual aid to our neighboring
towns and their responses to the Town of Farmington when needed.
Without mutual aid, no fire department would function effectively
today, work schedules, availability, and lack of commitment being
such as they are. The cascade system filled a total of 734 air
bottles this year. A new control panel was added to the cascade
system in the station to increase the efficiency when filling air
bottles.

Under the direction of Deputy Chief Clyde Ross, the fire
prevention and training programs were very busy as indicated in
his annual training report.

The requirements and responsibilities of the Fire Department are
on the increase. Annual inspections from OSHA and the Department
of Labor Standards require records on training, purchase of
turnout gear, medical exams, SCBA maintenance, Hepatitis 'B'
vaccinations, as well as our own vehicle maintenance
requirements.

A truck committee made up of several members of the Fire
Department has been busy writing specs for a new pumper truck.

Once again, on behalf of the Farmington Fire Department, I wish
to thank all those who have assisted us this year, especially the
Police Department, Highway Department, the personnel of Delta
Ambulance, the dispatchers at the County Dispatch Center and the
to the Ladies Auxiliary for providing food and beverages in times
of need.

Respectfully submitted,

Robert L. McCleery
Fire Chief

1994 Report of Fire Chief & Training Report

TRAINING REPORT FOR THE FIRE DEPARTMENT 1994

To: Chief McCleery and the Citizens of the Town of Farmington:

During 1994, the Fire Department has held 19 training sessions to comply with the current department SOP and the mandates of the State and Federal Agencies that govern emergency services. Some of the common topics that we refresh with each year will just go down as done. Of course, there are always the new regulations that have to be dealt with as they arrive. Some of the new training this year include the following topics:

To keep abreast of the current demands, we had to have sessions on Blood Borne Pathogens, Sexual Harassment in the work place, Cold Water Rescue, Confined Space Entry, and Rescue and Crime and the Fire Scene. Several of these courses were taught by local citizens and other professionals in our area. Added to these courses for the first time were Rope Rescues, Search and Rescue with the Maine Wardens Service, and Below Grade Rescue. Now you say, why?

Some of you citizens realize that in this day and age with people doing the many activities that they feel are necessary, they also find themselves in situations that are beyond their skills. When they need assistance, who do they call? You know that one of those agencies is the Fire Department because of the number of people who respond and the quickness of the response. The incident doesn't always meet the expectations of the citizen's idea of what a fire department should be doing. Do you want us to stop answering calls for help? Again, think of what you do and your skills.

This year has been a busy one with Fire Prevention and Education. Today we are beginning to see the benefit of years prevention programs have been taught. As you look at the Chief's report, you will see that calls are down in the Fire Emergency calls. Some people will say that it is just a sign of the times, but I feel that over the years our prevention education has been accepted by the young people and their families. We really appreciate the support and cooperation that has made these programs possible. We can thank the local insurance agencies and the school departments for their continued involvements. In our local schools, grades 1 to 3 are using the Learn Not To Burn Program and the teachers are the prime movers in doing the 25 different lesson plans each year. We also visit, or are visited at the Fire Station, the pre-schools and nursery schools that have an interest in the safety education of their children. This year we were in direct contact with some 560 children in the above mentioned programs. This is a wonderful effort by citizens to become involved. Thanks for your concern and support.

Training is on-going. We must maintain our current level of skills and keep an open mind for the new requirements and techniques. Some of you think this is all a waste of time and money; please reconsider your thoughts and come to the station someday and discuss your feelings and see what is expected of fire fighters today.

In closing, Fire Prevention Education and Fire Fighter Training are one of the constants that this Department is strongly committed to and will attempt to provide to all who need these services. I want to thank the citizens for their understanding and cooperation in the past year.

Sincerely yours,

S. Clyde Ross
Deputy Chief
Training Coordinator

1994

REPORT FROM THE FIRE DEPARTMENT

To: The Town Manager, Selectmen, and Citizens of Farmington:

This year the Fire Department was involved in the annual 4th of
July parade and fireworks, Easter Egg hunt, Children's Task Force
at the University of Maine -Farmington, Moonlight Madness Chicken
Bar-B-Q, annual Franklin County Fair and Chester Greenwood Day
Parade.

Under the direction of Deputy Chief S. Clyde Ross, the fire
prevention and training programs were very active as indicated in
his annual report. After much discussion and deliberation, the
truck committee selected an Emergency One Pumper Truck sold by
Fire Tech and Safety of New England. This new truck was
delivered to the Fire Department the first of October and after
passing the acceptance test and extensive training, the new "E
One" truck was put in service.

The Fire Department responded to 282 calls in 1995, 240
emergency and 20 service; also 22 training sessions. Some of the
emergency calls responded to were: 15 structure fires, 13
chimney fires, 10 vehicle fires, 75 accidents and extrication
calls, 10 grass, brush and debris fires, 22 false or alarm
malfunctions, 42 smoke, stove and electrical related fires, 14
gasoline and oil spills, 9 miscellaneous calls. Also mutual aid
to our neighboring towns and their response to the Town of
Farmington when needed.

Without mutual aid, no fire department would function effectively
today, work schedules, availability and lack of commitment being
such as they are. The cascade system filled a total of 300 air
bottles used in fire related incidents this year.

Once again, on behalf of the Farmington Fire Department, I wish
to thank all those who have assisted us this year, especially the
Police Department, Highway Department, the dispatchers at the
County Dispatch Center and to the Ladies Auxiliary for providing
food and beverages in times of need. Also to the personnel of
the former Delta Ambulance for their assistance. The Farmington
Fire Department expects to give and receive the same level of
cooperation with the new LifeStar Ambulance service as in the
past.

Respectfully submitted,

Robert L. McCleery
Fire Chief

1995 Report of Fire Chief & Training Report

TRAINING REPORT FOR THE FIRE DEPARTMENT 1995

To: Chief McCleery and the Citizens of the Town of Farmington:

Training and knowledge of one's responsibilities is a never ending process. The fire service is no exception. As you, the citizens, find yourselves in more complex situations, you require more of emergency services. Therefore, it is necessary that we continually update our skills.

Today, there are many new techniques and methods to be used in dealing with emergency situations. We read and hear every day about devastating incidents that claim lives and destroy properties. Hopefully, these will not come to our community. However, we must be as adequately prepared as possible. Training is one of the ways we prepare.

During the past year, there were numerous training and study sessions. Among these are those required by mandates (federal or state) and our local skill builders. Typical sessions center around HIV and Blood Borne Pathogens, Harassment in the work place, Pumps and Relay Techniques, ladder uses and maintenance, forcible entry, ventilation, auto extrication, low angle rescue, cold water rescue, CPR refresher and EMS assistance, operation of emergency vehicles, tanker operations, SCBA update and maintenance of equipment, and hoses and appliances.

The Fire Department has continued its prevention and education programs by assisting at the local elementary school, nursery and day care facilities, as well as other social service organizations. The response and participation in these groups are outstanding. Our thanks to them for all the help and assistance they offer in prevention.

Several of our fire fighters have attended the Fire Fighter's Academies offered by the State Fire Service Education Association, county or regional training sessions and those conducted locally with mutual aid departments. Some of our people have taken part as instructors and students.

This past summer, I had the opportunity to attend four fire seminars in Marlborough, Massachusetts. These were presented by leading professionals in the New England area; all were very informative and offered suggestions that have been implemented locally. These fire personnel have shared their ideas within our department.

Sincerely yours,

S. Clyde Ross
Deputy Chief
Training Coordinator

1995

To: The Town Manager, Selectmen, and Citizens of Farmington:

The personnel of the Farmington Fire Department were involved in
several activities this year, namely Chester Greenwood Day
Parade, the Easter Egg Hunt, Children's Task Force at the
University of Maine-Farmington, 4th of July Parade and Fireworks,
Moonlight Madness Chicken Bar-B-Q, the Annual Parade at
Farmington Falls, and Franklin County Fair.   Several firemen
attended the Maine State Federation of Firefighters Convention in
Ellsworth.   Chief McCleery and Deputy Chief Ross attended
meetings of the Maine Fire Chief's Association, the New England
Association of Fire Chiefs and the International Association of
Fire Rescue. Under the direction of Deputy Chief S. Clyde Ross,
the fire prevention, education and training programs were very
successful as indicated in his annual report.

In 1996, the Fire Department responded to 320 calls, 272
emergencies, and 20 service. Emergency calls responded to were:
9 structure fires, 13 chimney fires, 11 vehicle fires, 95
accidents and 15 extrication calls, grass, brush and debris
fires, 17 false or alarm malfunctions, 24 smoke, stove, furnaces
and electrical related fires, 29 gasoline, oil and hazardous
spills, 25 miscellaneous calls. Also mutual aid was given and
received from our neighboring towns when needed.

Without mutual aid, no fire department could function effectively
today, manpower availability, work schedules and lack of
commitment by members of volunteer or call fire departments.

On behalf of the Farmington Fire Department, I wish to thank the
Police Department, Highway Department, the dispatchers at the
county Dispatch Center, the Ladies Auxiliary for providing food
and beverages when needed, also to the personnel of Lifestar
Ambulance for their assistance, and the Town Manager, elected and
appointed officials for their support.

Respectfully submitted,

Robert McCleery
Fire Chief

1996 Report of Fire Chief & Training Report

TRAINING REPORT FOR THE FIRE DEPARTMENT 1996

To: Chief McCleery and the Citizens of the Town of Farmington:

The past year has seen an increase number of emergency training
sessions. The reasons for this has been the continuing calls
from the public to aid or assist with particular situations. We
are receiving more non-fire related calls than in the past.

Topics covered in fire training included the obvious fire related
schools with pumping water, raising ladders, S.C.A.B.,
ventilation, building construction, apparatus placement and
maintenance of the equipment. The mandated blood borne pathogen
class, physical examinations and sexual harassment updates were
also included. To this list one might add, hazardous materials,
electrical safe guards, propane situations, farm medics
situations (heavy farm equipment accidents) and hours of
buildings with special occupancies or problems.

Search and Rescue has been an other area that has received much
attention this year. Gary Anderson from the Department of Inland
Fisheries and Wildlife spoke about how we could network with
other units in the State to help better serve citizens in need.
High/Low angle rescue, ice, cold water rescue, and working with
an air boat from a neighboring community. This area of training
will continue to be of significance as more people use our
wilderness areas for vacation.

Fire prevention and education again this year received
considerable attention. More than 1200 students, K-8, were
involved with classroom discussion and school fire drills. Even
with this effort, there is a growing number of youths involved
with fire experimentations and our community has seen its share
of problems. In December, I participated in a two-day training
program at E.M.V.T.C. in Bangor. There were five other persons
from the area there also. We all need to take a greater
responsibility in teaching our young people how to be safe from
fire in our homes and businesses.

In closing, I wish to thank the Citizens for their continued
support and understanding of our programs. Thanks also to the
SAD #9 teachers who help with Fire Prevention and to the Town
Officers for their interest and concerns.

Respectfully submitted,

S.Clyde Ross
Deputy Chief
Training Coordinator

1996

REPORT FROM THE FIRE DEPARTMENT

To: The Town Manager, Selectmen, and Citizens of Farmington:

This year the Fire Department was involved in several activities,
namely, the Chester Greenwood Day Parade, the Easter Egg Hunt,
Children's Task Force at the University of Maine-Farmington, 4th
of July Parade and Fireworks, Moonlight Madness Chicken Bar-B-Q,
the annual parade at Farmington Falls, the Old Home Day Parade at
Phillips and the Franklin County Fair. Several firemen attended
the Maine State Federation of Firefighters Convention in
Kennebunk, Maine. Chief McCleery and Deputy Chief Ross attended
meetings of the Maine Fire Chief's Association, the New England
Division of the International Association of Fire-Rescue and New
England Association of Fire Chief's combined, at Springfield,
Massachusetts. Under the direction of Deputy Chief Ross, the
fire prevention education and training programs were very
successful as indicated in his annual report.

In 1997, the Fire Department responded to 252 calls, 208 were
emergency and 27 service calls. Emergency calls responded to
were 16 structure fires, 12 chimney fires, 12 vehicle fires, 65
accidents and extrication calls, 8 grass, brush and debris fires,
19 false or alarm malfunctions, 27 smoke, stove, furnace and
electrical related fires, 14 gasoline, oil and hazardous spills,
35 miscellaneous calls. Also, mutual aid was given and received
from our neighboring departments when needed. Without mutual
aid, no Fire Department could function effectively today. Work
schedules dictate the availability of manpower, especially during
the work week.

In the near future, the Fire Department should consider replacing
the rescue truck with a new heavy duty rescue truck to carry all
the rescue equipment that is now divided between two vehicles.
The present rescue truck is very overloaded.

On behalf of the Farmington Fire Department, I wish to thank the
Police Department, Highway Department, the dispatchers at the
County Dispatch Center, the Ladies Auxiliary for providing food
and beverages when needed, also the personnel of LifeStar
Ambulance for their assistance, and the Town Manager and elected
and appointed officials for their support.

Respectfully submitted,
Robert L. McCleery
Fire Chief

1997 Report of Fire Chief & Training Report

TRAINING REPORT FOR THE FIRE DEPARTMENT 1997

To: Chief McCleery and the Citizens of the Town of Farmington:

As another year closes, we find the Fire Department continuing
its busy training programs. Most of you have read these reports
year after year and wonder if we will ever change the topics of
the training schedules. As the needs of citizens change, so do
the demands made upon Emergency Services. It is necessary that
we continue our Fire Training with new skill development.

An example of these new requirements is that many more citizens
are taking to the outdoors today with new recreation and leisure
time activities. We are receiving more calls for assistance from
persons who are lost or injured in the wilderness areas of our
region. Hiking, climbing that mountain trail we always wanted to
experience, canoeing down the river or stream or skiing into the
back country. People need help and we have developed a Search
and Rescue Unit, assisted by other fire departments, to come to
the aid of injured citizens. We have also been called by the
Maine Wardens Service to assist them in locating and helping
injured hikers.

Fire fighting skills have to be maintained and new techniques
developed to meet the demands from the public. With the use of
more synthetic materials in construction and located in the
business community, we have to be prepared to face the challenges
from these items Hazardous materials are around us every day
and one needs to recognize the amount of these items that travel
by motor vehicles throughout our community each day. We have few
incidences but have to be ready to answer the calls.

This year we have continued with training in the topic areas that
you will find in past reports, no need to repeat them now. Town
firefighters have achieved the level of FF1 through the Maine
Fire Academy system (one on campus and one through local
delivery). This is in keeping with the current standards that
the department maintains.

Public Education is ongoing for the Fire Department. We
participated in the Fire Prevention Week activities at the local
elementary schools. This year 425 students in grades one through
three either visited the Fire Station or had a fire fighter visit
their classroom for safety presentations. Fire drills were
conducted in all the schools complying with standards established
at the State level. We must thank the educators for keeping our
youth aware of changes in today's society and how to cope with
emergencies. Thanks to a local insurance agent, we had the use

of a 911-telephone simulator in the elementary grades this past
Spring. The response of the children was outstanding. The Fire
Department has visited 6 nursery school units with personnel and
fire apparatus. There is a big difference in how children react
to this practice from one year to the next. Maturity and good
skills gained in school and at home have made this possible.
Thanks to all who have participated in these visits.

Farmington Fire Department is fortunate to have the assistance of
guest instructors as well as its own in-house personnel conduct
fire training programs for its members and mutual aid
departments. We have many qualified and capable people in the
fire service today, thanks to your support of our continuing
programs. We are making every effort to comply with the
standards established by the governing bodies that oversee the
fire services today.

Lastly, thanks to all of you who have contributed your time and
expertise to assist the Fire Department with its activities and
programs for our youth.

Sincerely yours,
S. Clyde Ross
Deputy Chief
Training Coordinator

1997

Robert L. McCleery 303

FIRE DEPARTMENT

To: The Board of Selectmen, Town Manager and Citizens of Farmington:

The year 1998 started off with the Ice Storm. The power was out for days, and weeks in some places; trees and power lines were down throughout the Town. Firemen assisted with the removal of trees from road ways and power lines and helped assist needy citizens to the Community Center shelter. Some firemen stayed at the shelter and helped with the food preparation. Firemen also hauled water with an engine to farms to water livestock. Combined with other calls, January was an eventful month with a total of 53 calls.

June was another busy month due to the heavy rain and flooding. Extensive damage was done to roads, fields and crop land. The Firemen patrolled flooded roads and streets in and around Farmington.

July brought heavy lightning strikes. The MTE, Inc. fire in Fairbanks on July 18th, which was the worst fire in Farmington in several years, involved several mutual aid departments.

The year 1998 saw four major fires: Bouffard's Warehouse on North Main Street in March; the MTE, Inc. fire; the Haley house fire on Knowlton Corner Road; and in November, Butterfly McQueen's on Broadway.

Chief McCleery and Deputy Chief Ross attended meetings of the Maine Fire Chief Association and the International Association of Fire Chiefs, New England Division in Springfield, Mass. As you read in Deputy Chief Ross' report, the fire prevention, education and training programs are very successful.

The Fire Department was involved in many community activities, namely, Children's Task Force at UMF, the Fire Department's annual Easter Egg Hunt, Fourth of July Parade and Fireworks, the annual parade at Farmington Falls, the Old Home Day Parade at Phillips, Moonlight Madness Chicken Bar-B-Q, Franklin County Fair and the annual Chester Greenwood Day Parade.

In August 1998 two young firemen completed the Fire Fighter II course held in Waterville. The nine days and evenings of study and field training is conducted by Maine Fire Service Training and Education.

In 1998 the Fire Department responded to 288 calls, 260 of these were emergency calls which included structure fires, vehicle fires, accidents and extrication calls, grass, brush and debris fires, false alarms or alarm malfunctions; smoke, stove, furnace and electrical related calls; gasoline, oil and hazardous spills, mutual aid requests from other Fire Departments, and search and rescue.

I wish to personally thank those Fire Departments that so generously responded to our calls for mutual aid. Without mutual aid, no Fire Department could function effectively today. Work schedules and manpower shortages make mutual aid an important part of all major emergencies.

Farmington Fire Department's Search and Rescue Team was called twice by the Warden's Service to assist with the rescue of injured climbers on Tumbledown Mountain in Weld. Farmington Fire Department's truck committee is considering future replacement of the rescue truck with a new heavy duty truck that will carry all of the rescue equipment in one vehicle.

On behalf of the Farmington Fire Department, I wish to thank the Police Department, Highway Department, the dispatchers at the County Dispatch Center, the Ladies Auxiliary for providing food and beverages when needed, also the personnel of Lifestar Ambulance for their assistance, and the Town Manager and elected and appointed officials for their support.

Respectfully submitted,

Robert L. McCleery
Fire Chief

FARMINGTON FIRE DEPARTMENT

Robert L. McCleery, Chief

S. Clyde Rust, Deputy Chief	Stephan M. Bunker
Terry S. Bell, Deputy Chief	Richard A. Cabot
Timothy A. Hardy, Asst. Chief	Jonathan E. Davis
John O. Bell, Captain	Jason A. Decker
Morrill W. Collins, Captain	David M. Frank
Harold H. Hemingway, Captain	Marc Hand
Stanley Wheeler, Chaplain	Timothy D. Hardy
George L. Barker, Lieutenant	Rocky Jackson
Michael A. Bell, Lieutenant	Curtis C. Lawrence
Peter H. Brennick, Lieutenant	Robert E. McCully
Nelson E. Collins, Lieutenant	Eugene L. Mosher
Richard A. Knight, Lieutenant	Douglas R. Oliver
Clyde H. Meader, Lieutenant	Raymond G. Pillsbury
Chet C. Alexander	David A. Portle
Jonathan M. Alexander	Donald A. Richard
Philip R. Allen	Gregory E. Rouz
Melvin L. Bard	Randall A. Voter
James A. Brown	Junior E. Turner

1998 Report of Fire Chief

~FIRE DEPARTMENT~

To: The Board of Selectmen, Town Manager and Citizens of Farmington

I submit my 23rd Annual Report of the Farmington Fire Department for the year ending December 31, 1999, the end of the millennium.

The Fire Department was involved in many community activities, namely the Children's Task Force at U.M.F., the Fire Department's annual Easter Egg Hunt, the Fourth of July Parade and Fireworks, the annual parade at Farmington Falls, the Old Home Days Parade at Phillips, Moonlight Madness Chicken Barbecue, Franklin County Fair and the annual Chester Greenwood Parade.

In August, the Fire Department sent two men to the Fire Fighter II Academy in Auburn. This is a nine-day and evening study and field training conducted by Maine Fire Training and Education.

In 1999, Farmington Fire Department responded to 274 calls; 237 of the calls were emergencies, which included structure fires, vehicle fires, accidents and extrication calls, hazardous spills, false alarms, mutual aide calls, furnace calls, debris fires, electrical-related fires and rescue calls. There were two structure fires resulting in total loss; the Brackett House in West Farmington in February, and the Abbott House in June, now the location of M.B.N.A.

The Fire Department's training program is continually on-going as is stated in Deputy Chief Ross's training report. These sessions and programs are necessary to keep abreast of changing laws and mandates facing Fire Services today. Today's Fire Departments are no longer just a Fire Department, they are Fire, Rescue and EMS combined into an Emergency Service Department.

This year an exhaust system has been installed in the apparatus bay to exhaust the toxic fumes from the trucks when they are started. This system will make the air cleaner providing a healthier work place for the whole building.

On behalf of the Farmington Fire Department, I wish to thank the Police Department, Highway Department, the dispatchers at the Dispatch Center, the Ladies Auxiliary for providing food and beverage when needed, also the personnel of Life Star Ambulance for their assistance, and the Town Manager, Selectmen and the ladies in the office for their support.

Respectfully submitted,

Robert L. McCleery
Fire Chief

1999 Chief Robert McCleery Last Annual Report

~Fire Department Training Report~

To: The Board of Selectmen, Town Manager, and Citizens of Farmington

As another year comes to an end, this report completes a decade of Fire Dept. Training. It is our practice to continue to upgrade the members in their skill levels and introduce new techniques and methods for meeting the emergencies from the public. The training schedule is an extensive list of topics, many of these have been in previous Town Reports and really don't need to be listed here.

We continue to meet the mandated items as presented in the statutes and new pieces of legislation. There are items on the legislative agenda this year which will have an impact on the fire service in the years to come; namely, the new Respiratory Statute we are trying to initiate. This will require more training sessions and improved guidelines.

Fire Prevention Week once again saw the Department working with the local elementary school children and the annual visit of the first graders to the Fire Station. Fire drills were held at the Mallet School and once again these youngsters and their teachers should be commended for the orderly evacuation of the building. Education is the key to reducing fire related injuries and property damage. Students had an opportunity to ask questions about fire safety when fire fighters visited their classrooms. The use of the 9-1-1 telephone simulator (made available by a local insurance agency) continued to raise awareness about where people live and how to report an emergency situation.

Several Nursery Schools and Day Care Centers were visited and in some cases, training sessions were conducted for the staff members. It is very important that care providers be aware of safety issues and how to address getting problem areas corrected. Cooperation of staff and board members is the only way this can be accomplished.

Fire Extinguisher classes were conducted at the high school for administrators and staff members. Food lab students learned about extinguishers and then discharged one of them. For some, this was a unique experience. The Fire Department continues to tour newly constructed buildings and frequently checks on the progress of new construction.

This year two of our members participated in the Fire Fighter II Academy in Auburn and others attended one and two day classes at the regional training sessions in different counties. Several officers have attended State meetings for new regulations and ideas being introduced into the emergency services.

Our members are active with the local Emergency Preparedness Committee and attend Hazardous Materials (Haz Mat) training and drills as part of the mutual aid response team. This is very important to our area as there are a wide variety of hazardous materials transported through our community each day.

In closing, I wish to thank all who have participated in our training through instruction, providing training locations, funding and understanding the need of a continuous educational process.

Respectfully submitted,

S. Clyde Ross
Dep. Chief and Training Coordinator

FARMINGTON FIRE DEPARTMENT

Robert L. McCleery, Chief

S. Clyde Ross, Deputy Chief	Stephen M. Bunker
Terry S. Bell, Deputy Chief	Richard A. Chabot
Timothy A. Hardy, Asst. Chief	Jonathan E. Davis
John O. Bell, Captain	Jason A. Decker
Morrill W. Collins, Captain	David M. Fronk
Harold H. Hemingway, Captain	Marc Hand
Stanley Wheeler, Chaplain	Timothy D. Hardy
George L. Barker, Lieutenant	Rocky Jackson
Michael A. Bell, Lieutenant	Curtis C. Lawrence
Peter H. Brannick, Lieutenant	Robert E. McCully
Nelson E. Collins, Lieutenant	Eugene L. Mosher
Richard A. Knight, Lieutenant	Douglas R. Oliver
Clyde H. Meader, Lieutenant	Raymond G. Pillsbury
Chet C. Alexander	David A. Pottle
Jonathan M. Alexander	Donald A. Richard
Philip R. Allen	Gregory E. Roux
Melvin L. Bard	Junior E. Turner
James A. Brown	Randall A. Voter

1999 Training Report

Rookie 1977

Deputy Chief 1999

Chief Terry S. Bell, Sr.
Farmington Fire Department's
First Full Time Fire Chief

2000-Present

FIRE DEPARTMENT

To: Board of Selectmen, Town Manager and Citizens of Farmington

I hereby submit my first annual report, for the Farmington Fire Department, for the year 2000. As this was the first half-year of having a full-time Fire Chief, the last 6 months have been very busy. The Department responded to 255 calls for assistance. Of those calls, 250 were emergencies. They included structure fires, chimney fires, vehicle fires, accidents and extrication calls. We also responded to alarm malfunctions, hazardous materials calls, gasoline and oil spills, and several miscellaneous calls. The other five were service calls.

A truck committee made up of members of the Fire Department is busy writing specs for a new pumper / rescue truck. This truck will replace Engine 5 that is in the Falls Station and the tanker that has gone to the Highway Department. We will also be moving some equipment from the squad truck to this new unit.

CHIEF TERRY BELL WITH NEW FIRE TRUCK

2000

I would like to personally thank Chief McCleery for all his help over the years to the Fire Department and myself. I would also like to personally thank all the members of the Farmington Fire Department for the fine job and help they have given to the Department. **"THANK YOU"**

I must thank the Mutual Aid Departments, which include Industry, Chesterville, Strong, New Vineyard, Temple, Wilton, New Sharon, and Jay, for all the help they have given to us over this past year. **"THANK YOU"**

I would like to take this opportunity to thank the members of the Ladies Auxiliary, Farmington Police Department, Highway Department and the ladies in the front office for all their help in the past year. **"THANK YOU"**

The Farmington Fire Department/Farmington Fire Department Benevolent Association would like to take this opportunity to **THANK** all the citizens, business groups and organizations who so generously supported the fund raising to purchase the Thermal Imaging Camera. The support from each of you made it possible to buy a state-of-the-art unit with necessary accessories.

This new Thermal Imaging Camera will be a valuable tool in helping to locate hidden fires and hot spots in buildings, persons inside a structure, and high heat sources in buildings during the overhaul mode. Training in the uses of this camera have already started and further sessions are scheduled.

Once again the Fire Department and Benevolent Association thank all who assisted with their contributions. Everyone should be proud of the total effort of the citizens.

Respectfully submitted

Terry S. Bell, Sr.
Fire Chief

FIRE RESCUE

To the Citizens, the Board of Selectmen, and the Town Manager:

I submit my annual report, for the Farmington Fire Rescue for the year 2001. The department has responded to 223 calls for assistance as of Dec. 6, 2001. This is how some of the calls are broken down. There have been 18 structure fires, 7 chimney fires, 9 car fires, 14 outside fires, 94 rescues/extrications, 10 service calls, 28 good intent, and 34 false alarms. The department has responded for mutual aid 27 times. Our members have attended 27 training sessions which required 78 hours to complete.

The Truck Committee has been meeting throughout the past three years and has put together the truck specifications for the pumper/rescue truck. We will be sending these specs out by the middle of December, 2001. This truck will be replacing Engine 5.

Farmington Fire Rescue is hosting the first Firefighter 1 program for high school students in Maine. I would like to recognize three of the many people who have been involved with this project, Fred Hyde, Wilton Fire Chief; Jack Berry, Maine Fire Service Training Instructor, and Ann Deraspe, Director of Foster Regional Applied Technology Center. These students come from Foster Regional Applied Technology Center. The class is made up of thirteen high school students and three adults. When these students become eighteen years old and have completed all the classes, they will be tested by the State of Maine. After successfully completing their State tests they will be certified at the level of Firefighter 1. This should be a big help to the area departments.

The department hosted an Instructor 1 class consisting of department members from around the county. As instructors they will be able to go anywhere in the state and help instruct firefighter classes and will be a big asset to their own departments. Four of these new instructors are members of our department.

Thank you, George Barker, Raymond Pillsbury and Curtis Lawrence for many years of service with the Farmington Fire Rescue.

2001

I would like to take this opportunity to thank the mutual aid departments which include Chesterville, Industry, Jay, Livermore Falls, New Sharon, New Vineyard, Strong, Temple and Wilton and their respective Auxiliary members; the International Paper Haz-Mat team, and the Franklin County EMA office.

I would also like to thank our Ladies' Auxiliary, Police and Public Works Departments; Franklin County Sheriff's Department, their dispatchers; and the members of LifeStar Ambulance. A very special **"THANK YOU"** to the officers and the members of our department.

Please visit our website at www.farmingtonfirerescue.org.

Respectfully submitted,

Terry S. Bell, Sr.
Fire Chief

Farmington Fire Rescue Department Members
2000-2014
#435 Jennings Pinkham Above Left

Photos Courtesy Town of Farmington

under the charter of the Farmington Village Corporation approved by[?] Governor, March 30, 1911 and adopted by the Farmington Village Corporation, January 19, 1912, an act to organize and maintain an efficient Fire Department and to adopt all rules and regulation for governing the same. (of chapter 102, Private and Special Laws of 1911.)

The name and title of this association shall be the Farmington Fire Company, No. 1.

The members shall be residents of the Farmington Village corp. and of good moral character and temperate habits

The officers of the fire company shall consist of a Foreman who shall be the Chief Engineer of the department and appointed by the Assessors. Other officers were first and Second assistant Foremen, Clerk, Treasurer and a Finance Committee of three, one of whom shall be the first assistant Foreman. These officers shall be elected by ballot at the annual meeting held on the first Tuesday of December. The company shall hold regular monthly

Establishment and Growth — Narrative

ORIGINAL COPY The Chronology

A Brief History of Farmington Fire Department

The first record of any fire in the town of Farmington appears in Butler History of Farmington in 1850, destroying the square between Main St., Broadway and Exchange St. At this time there was no organized Fire Protection in the Town and some of the citizens began inquiring into the possibility of organizing a Fire Department but nothing came of it. From the Franklin Patriot Jan. 1859 an act of Incorporation accepted for the Farmington Village corporation establishing regulations for a Fire and Police Departments. Nov. 16, 1860 a fire Engine purchased a Hand engine company organized to work it. From the Farmington Chronicle dated Thur. Nov. 29, 1860 I quote "Fire! By the train of Friday last came the much talked of Fire Engine for the use of Farmington Center Village, and on Saturday afternoon the fire company, under the direction of "Captain Jennings", took possession of her merchism[?] at the Depot, and two abreast escorted it to the village in fine style. Passing up Main Street, the "boys" greeted the officers of the corporation, assembled on the line of march, with three rousing cheers, and passed on to the court House reservoir, where the hose was run off, the brakes manned, and a trial of "Eagle I," exhibited to the village audience, free of charge, and pronounced